Place value

1 (a) Write the number twenty-five thousand, three hundred and two in figures.

   2.....................................   **(1 mark)**

   (b) Write the number 12 317 in words.

   Twelve thousand, ...   **(1 mark)**

   (c) Write down the value of the 7 in the number 327 332.

   7.....................................   **(1 mark)**

2 Complete the place value table showing nineteen thousand and sixty-one.

10 000s	....	....	10s	....
1	....	....	....	1

     (2 marks)

3 Write the following numbers in order starting with the lowest number.

   (a) 251, 209, 219, 199, 211

   199, ...   **(1 mark)**

   (b) 3711, 3910, 3099, 3118, 3796

   3099, ...   **(1 mark)**

   > The 'thousands' are all the same.
   > Look at the values of the 'hundreds'.

4 Write the measurements in order of size starting with the smallest.

   (a) 31 851 km, 31 809 km, 31 787 km, 29 738 km, 32 001 km

   ....   **(1 mark)**

   (b) 5.75 m, 5.09 m, 5.11 m, 5.8 m, 5.92 m

   ....   **(1 mark)**

5 The weekly pocket money of six children has been recorded. Which two amounts would you need to swap so that all the amounts are in order?
£6.10, £6.25, £6.70, £6.55, £6.35, £6.85

   ....   **(1 mark)**

6 A teacher has been given £35 to spend on pencils and rulers.
A packet of 12 pencils costs £2.15.
A pack of 30 rulers costs £6.00.
She buys 4 packs of rulers and spends the rest on packets of pencils.
How many packets of pencils can she buy?

   > You will need to use problem-solving skills throughout your exam – **be prepared!**

   PROBLEM SOLVED!

   ....   **(2 marks)**

Guided

Negative numbers

1 (a) Write the following numbers in order, smallest first.

$8, -5, -12, 0, 3$

....

> Use the fact that −12 is smaller than −5, and that any negative number is smaller than 0.

(1 mark)

(b) Solve the following.

(i) $-13 + 4$

$= -(13 - 4) = $ ....

> Remember that −7 + 3 = −(7 − 3).

(1 mark)

(ii) $-4 - -12$

$= -4 + $ .... $= $ ....

> Remember that −3 − −7 = −3 + 7 = 7 − 3.

(1 mark)

(iii) $-20 - 10$

$= -(20 + $ ....$) = $ ....

(1 mark)

(iv) $-15 - +7$

$- $ .... $- $ .... $= $ ....

> Remember that −3 − +7 = −10.

(1 mark)

2 Solve the following.

(a) $-15 - +7$

$= -15 - 7 = $ ....   **(1 mark)**

(b) $56 \div -8$

$= -(56 \div 8) = $ ....   **(1 mark)**

(c) -7×-7

$= 7 \times 7 = $ ....   **(1 mark)**

(d) $-64 \div -4 = $

....   **(1 mark)**

3 On a certain day in Helsinki, the temperature at noon was 6 °C. By 6 pm it had dropped by 8 °C. By 9 pm it had dropped a further 7 °C and by midnight it had dropped a further 7 °C. Find the temperatures at

> You can see each temperature drop as a subtraction of the number of degrees each time.

(a) (i) 6 pm

$6°C - 8°C = -2°C$ **(1 mark)**

(ii) 9 pm

$-2°C - 7°C$ ....°C   **(1 mark)**

(iii) midnight

............°C

(1 mark)

(b) What was the overall drop in temperature from noon to midnight?

............°C   **(1 mark)**

4 The following table lists the elevation (above/below sea level in metres) of cities around the world.

Prague	Tokyo	Amsterdam	Nairobi	Baku
244 m	17 m	−2 m	1728 m	−28 m

(a) What is the difference in elevation between Prague and Amsterdam?

....

(1 mark)

(b) Another city has an elevation halfway between that of Tokyo and Prague. What is its elevation?

....

(2 marks)

REVISE EDEXCEL GCSE (9–1)
Mathematics
Foundation

GUIDED REVISION WORKBOOK

Series Consultant: Harry Smith

Author: Russell Timmins

Notes from the publisher

While the publishers have made every attempt to ensure that advice on the qualification and its assessment is accurate, the official specification and associated assessment guidance materials are the only authoritative source of information and should always be referred to for definitive guidance.

Pearson examiners have not contributed to any sections in this resource relevant to examination papers for which they have responsibility.

For the full range of Pearson revision titles across KS2, KS3, GCSE, Functional Skills, AS/A Level and BTEC visit:
www.pearsonschools.co.uk/revise

Question difficulty
Look at this scale next to each exam-style question. It tells you how difficult the question is.

Contents

A small bit of small print

Edexcel publishes Sample Assessment Material and the
Specification on its website. This is the official content and this book
should be used in conjunction with it. The questions in this book
have been written to help you practise what you have learned in your
revision. Remember: the real exam questions may not look like this.

Rounding numbers

1 Round

  (a) 11 349 (nearest hundred)

   = 11............

| When rounding to the nearest 'hundred', look at the 'tens' column to decide if you should round 'up' or leave alone. |

(1 mark)

  (b) 13 459 (nearest thousand)

   = 1............

| Always go to the place value column to the right of the column to be rounded to make your decision. |

(1 mark)

  (c) 21 997 (nearest ten)

   = 2............

(1 mark)

2 Round 0.003 272 correct to

  (a) 1 significant figure

   ....

| The non-zero digits keep the same place value in your significant figures answer. The final one may increase by 1 (or stay the same) when it is rounded. |

(1 mark)

  (b) 3 significant figures

   ....

(1 mark)

  (c) 2 significant figures

   ....

(1 mark)

3 Round 361 712 correct to

  (a) 1 significant figure

   400 000

(1 mark)

  (b) 2 significant figures

   = 3 .... 0 000

(1 mark)

  (c) 3 significant figures

   ....

(1 mark)

4 Three athletes, A, B and C, recorded times for the 100 m sprint. Their times were 11.051, 10.923, and 11.114, respectively.

  (a) Round B to 3 significant figures.

   ....   **(1 mark)**

  (b) Round C to 2 significant figures.

   ....   **(1 mark)**

  (c) How many significant figures must they all be rounded to in order for the times to all be the same?

   ....

(1 mark)

5 Hans checked his bank account which said he had €342 617. He said he had €343 000 to 3 significant figures. Explain why you think he was either correct or incorrect.

| You will need to use problem-solving skills throughout your exam – **be prepared!** |

PROBLEM SOLVED!

   ....

   ....

   ....

(2 marks)

Adding and subtracting

1 Work out

(a) $556 + 19 + 375$

```
  556
   19
+375
.....0
    2
```

....

(1 mark)

(b) $804 - 257$

```
  7 9 14
  8̶0̶4̶
-257
.....7
```

....

(1 mark)

> Remember, when the sum of the digits exceeds 9 you carry the digit to the left place value.

2 Work out

(a) $4095 + 5863$

....

(1 mark)

(b) $9191 - 2658$

....

(1 mark)

3 A packet of pizza flour costs £1.52, 2 cans of tomatoes cost £0.58 each, 4 bags of mozzarella cheese cost £1.13 each and a packet of oregano costs £0.94. Alex pays with a £10 note. Work out how much change there will be.

$152 + (2 \times 58) + (4 \times 113) + 94 =$ ....

$1000 -$ .... $=$ ....

....

> Convert everything to pence, then convert back to pounds at the end.

(3 marks)

4 A bus leaves Guildford station with 63 passengers on board. At the next stop, 37 people get off and 15 others get on. How many passengers are on board now?

> Method 1: subtract the number of passengers who get off from the original 63 then add the number of passengers who get on.
> Method 2: subtract the 15 passengers from the 37 that get off, then subtract the result from the original 63.

....

(2 marks)

5 Manny needs to buy a saw for £3.24, a bag of nails for £1.61 and a hammer costing £5.38. He says a £10 note will be enough to get all three items. Is he correct? Explain your reasoning.

Guided

....

(3 marks)

Multiplying and dividing

1 Work out

(a) 35×42

The answer to 35×40 goes here. First write 0 in the units position, now work with the 4 (tens). Work out 4×5 (tens), write down the 0 in the tens position and carry the 2 (hundreds). Then work out 4×3 (hundreds), add on the 2 and write this result in the hundreds and thousands positions.

This can be done by multiplying 35×2, multiplying 35×40 and adding the two results together.

The answer to 35×2 goes here. Work out 2×5. Write the 0 from your answer in the units position and carry the 1. Then work out 2×3 (tens) and add the 1 to it. Write this digit in the tens position.

Finally, add your results together.

$$\begin{array}{r} 35 \\ \times\, 42 \\ \hline \dots\dots \\ \dots\dots \\ \dots\dots0 \\ \end{array}$$

....

(1 mark)

(b) $292 \div 4$

Reading the digits 292 from left to right gives three numbers 2, 29 and 292. The smallest that can be divided by 4 is 29. Work out how many 4s go into it: $7 \times 4 = 28$ giving a remainder of 1. Write the 7 (tens) on the top above the 9 (tens).

The remainder, 1 (ten), is written to the left of the units digit to make 12. Now you work out how many 4s go into 12. Write your answer in the units space on the top line.

$$4\overline{)2\,9^1\!2}\;\;\overset{7}{}$$

....

(1 mark)

2 Pens come in three packaging sizes: a packet, a bag and a box. A packet contains 8 pens, a bag contains 4 times more, and a box contains 3 times more than that.

(a) Calculate how many pens are in a box.

$= 8 \times 4 \times \dots\dots = \dots\dots$ **(1 mark)**

(b) Calculate how many pens you would have, if you had one of each packaging size.

$= 8 + \dots\dots + \dots\dots = \dots\dots$ **(3 marks)**

3 A farmer filled 26 boxes with carrots from a small field. Each box contained 27 kg of carrots. What was the total weight of carrots from his field?

$$\begin{array}{r} 26 \\ \times\, 27 \\ \hline {}_4 2 \\ \hline \\ \hline \end{array}$$

(3 marks)

 4 Work out

$354 \div 11$

Write any remainder as a fraction.

$$11\overline{)3\,5\,4}\;\;\overset{3}{}$$

Remember you are looking for how many 'groups' of 11 there are in each place holder so any remainders will be 11ths.

....   **(2 marks)**

 Guided

 PROBLEM SOLVED!

5 A lorry had a cargo of 600 bags of flour. Each bag weighed 8 kg. The driver delivered 240 bags to a supermarket and the rest was shared between 9 small shops.

You will need to use problem-solving skills throughout your exam – **be prepared!**

(a) What was the total weight of the flour in the lorry?

....   **(1 mark)**

(b) How many kg of flour did each smaller shop get?

....   **(2 marks)**

Decimals and place value

1 (a) Write down the value of 4 in 1.04

= 4.....................

> For example, in 1.27 the 2 represents the value 2 tenths.

(1 mark)

(b) Write down the value of 2 in 0.326

= 2.....................

> Each place-holder is ten times smaller every move to the right, so we have hundreds, tens, units, tenths, hundredths, etc.

(1 mark)

(c) Write down the value of 3 in 5.003

....

(1 mark)

2 Write the following numbers in order of size starting with the **smallest**.
8.3, 4.9, 6.7, 8.2, 7.6

....

(1 mark)

3 Write down the following numbers in order starting with the **largest**.
1.532, 1.499, 1.53, 1.6, 1.504

> Start by looking at the first place holder in every number to identify the largest digit, then move on to the second place holder, then the third and so on.

...............................1.499

(1 mark)

4 Write down the following numbers in order, **smallest** first.
0.81, 0.8, 0.08, 0.788, 0.019

....

(1 mark)

5 Use the information $1.2 \times 9.3 = 11.16$ to write down the value of

(a) 1.2×93

....

> This is exactly 10 times larger than 1.2 × 9.3.

(1 mark)

(b) 0.12×9.3

....

> 0.12 is 10 times smaller than 1.2 so the result should be 10 times smaller also.

(1 mark)

(c) 120×0.93

....

> 120 is 100 times larger than 1.2 and 0.93 is 10 times smaller than 9.3. If you multiply something by 100 then divide by 10 it is the same as just multiplying by 10.

(1 mark)

(d) $11.16 \div 0.12$

> From the fact you are given, 11.16 ÷ 1.2 = 9.3. Since 0.12 is 10 times smaller than 1.2, and you are dividing by this smaller number, the result will be 10 times bigger.

....

(1 mark)

6 Using the fact that $19 \times 21 = 399$, change **only one** of the numbers in the following to make it correct.
$39.9 \div 0.21 = 0.019$

> You will need to use problem-solving skills throughout your exam – **be prepared!**

PROBLEM SOLVED!

....

(2 marks)

Operations on decimals

1 (a) Work out 3.71 + 8.62

$$
\begin{array}{r}
3.71 \\
+\,8.62 \\
\hline
\text{........}3
\end{array}
$$

> Place value is crucial when adding or subtracting decimals, so keep them all in line, including the decimal point. If the sum of any column exceeds 10 then the 10 is carried to the left as a 1.

(1 mark)

(b) Work out 91.6 − 38.4

$$
\begin{array}{r}
91.6 \\
-\,38.4 \\
\hline
\text{.......}2
\end{array}
$$

> You may need to put some zeroes in any empty decimal places.

(1 mark)

(c) Work out 4.91 × 12

> You can calculate without a decimal point and then adjust the size of the answer. 4.91 is 100 times smaller than 491, therefore the result should be 100 times smaller than 491 × 12.

....

(2 marks)

(d) Work out 0.52 × 0.03

....

(2 marks)

(e) Work out 61.2 ÷ 6

> Think about 6 dividing 612 rather than 61.2 and then divide the result by 10. Or with the working shown to the right you can include the decimal point in both the working and the answer.

$$6\overline{)61.2}$$

....

(2 marks)

(f) Work out 45.1 ÷ 1.1

....

(2 marks)

2 A day ticket at an adventure park costs £33.60. A group of 6 people buy one each. What is the total cost of the tickets bought?

....

(2 marks)

3 Davina worked 38 hours last week and earned £577.60. How much does she earn an hour?

£..

(3 marks)

4 Carmen runs a music shop. She aims to sell 8 boxes of guitar strings every day from Monday to Friday. A box of strings costs £7.45. If she reaches her target, how much money will she have taken for strings in a 4 week period?

> You will need to use problem-solving skills throughout your exam – **be prepared!**

£.............................. **(4 marks)**

> 1. Work out how much she would take each day.
> 2. Multiply this by 5 to see how much is taken per week.
> 3. Multiply this by 4 to see how much is taken in 4 weeks.

Squares, cubes and roots

1 Work out

(a) 6^2

$6^2 = 6 \times 6 =$ .... **(1 mark)**

(b) 3^3

$3^3 = 3 \times 3 \times 3 =$ .... **(1 mark)**

(c) $\sqrt{36}$

| Work out which number squared is equal to 36. |

.... **(1 mark)**

(d) $\sqrt{144}$

.... **(1 mark)**

(e) $\sqrt[3]{27}$

| Work out which number cubed is equal to 27. |

.... **(1 mark)**

(f) $\sqrt[3]{64}$

.... **(1 mark)**

(g) $\sqrt[3]{125}$

.... **(1 mark)**

(h) $\sqrt[3]{-8}$

| Work out which number cubed is 8, and because $- \times - \times - = -$ you can then insert a minus sign. |

.... **(1 mark)**

(i) $\sqrt[3]{-27}$

.... **(1 mark)**

2 Work out the value of $4^2 - (2^2) - (3^2)$.

$(4 \times 4) - (2 \times 2) - (3 \times 3)$

.... – .... – .... = ...

.... **(1 mark)**

| Square the numbers first, then follow this with addition and subtraction. |

| You will need to use problem-solving skills throughout your exam – **be prepared!** |

PROBLEM SOLVED!

3 Look at the following numbers: 2, 4, 8, 9, 29, 27, 49, 64. Write down a number that

Guided

(a) is 1 less than a square number.

.... **(1 mark)**

(b) is 2 more than a cube number.

.... **(1 mark)**

(c) has a cube root of 4.

.... **(1 mark)**

4 If you add together four consecutive square numbers, the answer will always be an even number. Explain why you think this is either a true or false statement.

PROBLEM SOLVED!

| What happens when you square an even number? How about an odd number? |

....

....

.... **(2 marks)**

Indices

1 Write as single powers of 5

(a) $5 \times 5 \times 5$

(b) $5 \times 5 \times 5 \times 5 \times 5 \times 5$

$5 \cdots$ **(1 mark)**

$5 \cdots$ **(1 mark)**

2 Simplify the expressions, leaving answers in index form.

(a) $3^3 \times 3^4$

(b) $4^5 \times 4^3$

When multiplying with indices you add the indices together.

$\underbrace{3 \times 3 \times 3}_{3} \times \underbrace{3 \times 3 \times 3 \times 3}_{4} = 3^{3+4}$

$= 4^{5+3}$

.... **(1 mark)**

.... **(1 mark)**

(c) $5^2 \times 5^7$

(d) $3^5 \div 3^2$

When dividing numbers with indices you subtract the indices.

.... **(1 mark)**

.... **(1 mark)**

(e) $7^8 \div 7^2$

(f) $a^5 \div a^2$

$7 \cdots$ **(1 mark)**

.... **(1 mark)**

3 Write as a single power of 4

(a) $\dfrac{1}{4}$

(b) $\dfrac{1}{4 \times 4 \times 4}$

$\dfrac{4^1}{4^2} \quad \dfrac{\cancel{4}}{\cancel{4} \times 4} = \dfrac{1}{4} = 4^{1-2} = 4^{-1} \quad \boxed{4^{-1} = \dfrac{1}{4}}$

.... **(1 mark)**

.... **(1 mark)**

4 Simplify these expressions and leave your answers in index form.

 (a) $\dfrac{2^3 \times 2^3}{2^4}$

(b) $\dfrac{5^7}{5 \times 5^3}$

.... **(2 marks)**

.... **(2 marks)**

(c) $\dfrac{8^4 \times 8^3}{8^2 \times 8}$

(d) $\dfrac{3^{-2} \times 3^5}{3^4 \times 3^{-7}}$

.... **(2 marks)**

.... **(2 marks)**

5 Rewrite the following expressions without indices.

 (a) $33^0 = \dots$

(b) $25^{-1} = \dfrac{1}{\dots}$

Any number to the power 0 (except zero itself) is equal to 1.

(1 mark)

(1 mark)

(c) $\left(\dfrac{1}{2}\right)^3 = \dfrac{1^3}{2^3} = \dots$

(d) $\left(\dfrac{2}{3}\right)^{-3} = \left(\dfrac{3}{2}\right)^3 = \dots$

(1 mark)

(1 mark)

6 Work out the value of n.

 $3^5 \times 3^3 = \dfrac{3^{20}}{3^n \times 3^9}$

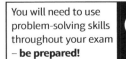
You will need to use problem-solving skills throughout your exam – **be prepared!**

.... **(2 marks)**

Estimation

 1 Give an estimated value of

(a) 87×103

> Round to the nearest whole single digit number or power of 10 to make estimated values easier to calculate.

$90 \times 100 =$ ....

(1 mark)

(b) $37.5 + 18.52$

....

(1 mark)

(c) $(19.42)^2$

$(.................)^2 =$ ....

(1 mark)

 2 Write an estimated value of $\dfrac{374.8}{9.7 \times 0.97}$

$\dfrac{374.8}{9.7 \times 0.97} = \dfrac{400}{10 \times ....} = \dfrac{.........}{.........} =$ ....

> Round all to 1 significant figure then calculate an estimation.

(2 marks)

 3 Write an estimated value of $\dfrac{17.53}{2.87 \times 6.42}$

$\dfrac{17.53}{2.87 \times 6.42} = \dfrac{.............}{3 \times ......} = \dfrac{.........}{.........} =$ ....

(2 marks)

 4 Write an estimated value of $\dfrac{4.65 \times 5.032}{0.48}$

$\dfrac{5 \times 5}{0.5} = \dfrac{.........}{.........} =$ ....

> You do not have to divide by a decimal. Simply multiply both numerator and denominator by 10 to get $5 \times 5 \times \frac{10}{5}$. Alternatively, you might notice that dividing by 0.5 (or $\frac{1}{2}$) is the same as multiplying by 2. So your calculation becomes $5 \times 5 \times 2$.

(2 marks)

 5 Give an estimated value of $\dfrac{173 \times 2.495}{0.783}$

....

(2 marks)

 6 Work out an estimated value of $\dfrac{6.013 \times 7.57}{0.21 \times 8.3}$

....

(2 marks)

 7 (a) Work out an estimated volume for a piece of rectangular cardboard measuring

$9.54 \text{ cm} \times 37.6 \text{ cm} \times 0.24 \text{ cm}$

$=$ .... cm^3

(2 marks)

(b) Give a reason why you think your estimate is either lower or higher than the true volume of the piece of cardboard.

....

....

(1 mark)

Estimation p. 1e

(1) C) $(19.42)^2$ why

change to 2e not 19

$$3) \frac{17.53}{2.87 \times 6.42} =$$

7) a) ?   why ~~205~~

0.24 → 0.25 ?

0 9119950251

- 25.2   Toys أَلْعَاب (2)
  glasses

      (11 & 12) رُسُومٌ

- explicit - implicit - convey
- extract - lifelike - impression

quotation - tailored - le

synoptic - lengthy - prompt

- picture prompt -
- incorporate - (cartoon) strip -

low - stakes - assignment.

-

Factors, multiples and primes

1 Write down all the factors of 24.

(a) (1 × 24), (2 × 12), (...... × ......), (...... × ......)

     = {1, 2, ..., ..., ..., ..., 12, 24}

> Remember, a factor of 24 is any number which exactly divides 24.

(2 marks)

(b) Write the next 6 multiples of 8.

     24, 32, 40, ..., ..., ..., ..., ..., ...

(1 mark)

2 Choose all the numbers in the box that are either multiples or factors of 14.

24	42	70
1	7	
64	49	

....

(2 marks)

3 Write the first 6 prime numbers.

     = ...2...   ...3...   ....

> Remember, a prime number p has 2 and only 2 factors which are always 1 and p itself.

(1 mark)

4 If all factors of 100 pence (£1) were minted as coins

(a) what would each coin be worth in pence?

     = 1, 2, 4, ...

> Find all the numbers that divide into 100 exactly.

(2 marks)

(b) how many different coins would be available up to and including £1?

     .... coins

(1 mark)

5 Danuta is thinking of a number between 40 and 50 that is the sum of three consecutive factors of 60 when the factors are ordered in terms of size.

(a) What are the three consecutive factors?

     ....

> Find all the factors of 60 first.

(2 marks)

(b) What number is she thinking of?

     ....

(1 mark)

6 Write out the following numbers as products of their prime factors.

(a) 84

     84 = 2 × 2 × 3 × 7 = 2^2 × 3 × 7

(1 mark)

(b) 40

     ....

(1 mark)

(c) 42

> Can you use the facts from (a) to help you with (c)?

     ....

(1 mark)

(d) 200

     ....

(1 mark)

HCF and LCM

1 (a) Write down highest common factor (HCF) of 60 and 75.

> Find all the factor pairs of both numbers. Use the highest factor that is in both lists.

60 = 1 × 60, 2 × 30, 3 × 20, 4 × ......, ... × ......, ... × ......
75 = 1 × 75, 3× ......, 5 × ......
HCF (60, 75) = 15

(2 marks)

(b) What is the lowest common multiple (LCM) of 12 and 15?

LCM (12, 15) =

> Start listing the multiples of 12 and the multiples of 15. The first number to appear in both lists is the LCM.

....

(2 marks)

2 Write the numbers 56 and 196 as products of their primes.

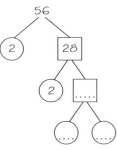

56 = ...

196 = ...

(4 marks)

3 Use the Venn diagram to find the LCM of 12 and 18.

12 = 2 × 2 × 3
18 = 2 × 3 × 3

> Find the prime factors of 12 and 18 first.

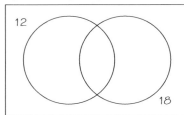

> The Venn diagram not only contains the HCF in the intersection but when this area is multiplied by the remaining two areas it produces the smallest possible multiple (LCM) of both numbers.

LCM (12, 18) = 2 × 3 × ... × ... = ...

(2 marks)

4 Find

(a) the HCF of 12 and 22

> One Venn diagram will be sufficient to solve both (a) and (b).

....

(2 marks)

(b) the LCM of 12 and 22

> You will need to use problem-solving skills throughout your exam – **be prepared!**

....

(1 mark)

Fractions

1 What fraction of the array is shaded?

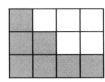

> The total number of squares in an array show how many equal parts make the whole. The shaded parts show a fraction of the whole.

$= \dfrac{\text{......}}{12}$ **(1 mark)**

2 Write the fractions in their lowest terms, then solve. Give your answers in their simplest form.

(a) $\dfrac{12}{18} - \dfrac{3}{9}$

$= \dfrac{2}{3} - \dfrac{1}{3}$

$= \dfrac{\text{......}}{\text{......}}$ **(1 mark)**

(b) $\dfrac{8}{30} + \dfrac{6}{45}$

$= \dfrac{\text{......}}{\text{......}}$ **(1 mark)**

(c) $\dfrac{14}{24} + \dfrac{15}{36}$

....   **(1 mark)**

(d) $\dfrac{3}{12} + \dfrac{4}{10}$

....   **(1 mark)**

3 Write the fraction of each array that is shaded, in its simplest form.

(a)

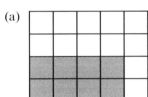

$= \dfrac{\text{......}}{20}$

$= \dfrac{\text{......}}{\text{......}}$ **(2 marks)**

(b)

$= \dfrac{\text{......}}{30}$

$= \dfrac{\text{......}}{\text{......}}$ **(2 marks)**

4 Work out

(a) $\dfrac{3}{8}$ of 160 km

> Draw an array consisting of 8 equal parts.

$\dfrac{3}{8} = \dfrac{3 \times 20}{8 \times 20} = \dfrac{60}{160}$

So $\dfrac{3}{8}$ of 160 km = ... km   **(1 mark)**

(b) $\dfrac{2}{7}$ of £56

£...........................   **(1 mark)**

(c) $\dfrac{9}{16}$ of 48 kg

.... kg   **(1 mark)**

5 Ella bought 24 boxes of melons each costing £9. Each box contained 12 melons. She sold $\dfrac{2}{3}$ of the melons for £1.20 each, then she sold the remaining melons for £0.90 each. How much profit did Ella make that day?

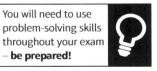

> You will need to use problem-solving skills throughout your exam – **be prepared!**

PROBLEM SOLVED!

£.....................   **(2 marks)**

Operations on fractions

 1 Work out

> To add or subtract fractions, you must first change the fractions to have a common denominator.

(a) $\frac{1}{3} + \frac{1}{6}$

$= \frac{1 \times 2}{3 \times 2} + \frac{1}{6} = \frac{2}{6} + \frac{1}{6}$

$= \frac{3}{6} = \frac{1}{2}$ **(2 marks)**

(b) $\frac{5}{6} - \frac{1}{4}$

$= \frac{\ldots}{12} - \frac{\ldots}{12} = \frac{\ldots}{12}$ **(2 marks)**

(c) $\frac{1}{5} + \frac{2}{7}$

$= \frac{\ldots}{35} + \frac{\ldots}{35} = \frac{\ldots}{35}$ **(2 marks)**

(d) $\frac{7}{9} - \frac{3}{4}$

$= \frac{\ldots}{\ldots} - \frac{\ldots}{\ldots} = \frac{\ldots}{\ldots}$ **(2 marks)**

 2 Work out

(a) $\frac{1}{4} \times \frac{1}{2}$

$= \frac{\ldots}{2 \times 4} = \frac{\ldots}{8}$ **(1 mark)**

(b) $\frac{2}{3} \times \frac{4}{5}$

$= \frac{\ldots}{3 \times 5} = \frac{\ldots}{15}$ **(1 mark)**

(c) $\frac{1}{5} \div \frac{3}{10}$

$= \frac{1}{5} \times \frac{10}{3} = \frac{\ldots}{\ldots}$

> Using the inverse operator (multiplication) and the inverse of the fraction that is dividing we can change the division to a multiplication, which is easier.

   (2 marks)

(d) $\frac{5}{8} \div \frac{15}{16}$

....   **(2 marks)**

 3 In a bag of fruit $\frac{1}{4}$ are bananas, $\frac{2}{5}$ are apples and the remainder are peaches. What fraction of the bag of fruit are peaches?

> Think of the bag of fruit as 'the whole'. If you add together the fractions representing the bananas and the apples, then subtract this from 1, you are left with the fraction of the bag that are peaches.

 You will need to use problem-solving skills throughout your exam – **be prepared!**

$= \frac{\ldots}{\ldots}$ **(3 marks)**

4 Vanessa shares £18 between her three children for pocket money each week. Marie gets $\frac{4}{9}$ of it and Paul receives $\frac{3}{8}$.

(a) What fraction of the pocket money is left for Lenny?

....   **(3 marks)**

 You will need to use problem-solving skills throughout your exam – **be prepared!**

(b) How much does Lenny get?

....   **(2 marks)**

Mixed numbers

1 Work out

(a) $2\frac{4}{5} + 4\frac{5}{6}$

> Add the proper fractions then add 2 + 4 to this result.

$= \frac{4}{5} + \frac{5}{6} + 2 + 4 = \frac{\dots}{5 \times 6} + \frac{\dots}{5 \times 6} + 6 = \dots\dots \frac{\dots}{\dots}$ **(2 marks)**

> Alternatively, convert mixed numbers to improper fractions, then convert back to proper fractions when the result is worked out.

$2\frac{4}{5} + 4\frac{5}{6} = \frac{14}{5} + \frac{29}{6} = \frac{14 \times 6}{30} + \frac{29 \times 5}{30} = \dots\dots + \frac{\dots}{30} = \dots\dots \frac{\dots}{\dots}$

(b) $5\frac{1}{3} - 2\frac{3}{7}$

> With subtraction problems it is always wise to convert to improper fractions, as the fraction part of the subtracted mixed number is often larger than the fraction in the first one.

$\frac{16}{\dots} - \frac{\dots}{7} = \frac{\dots}{21} - \frac{\dots}{\dots} = \frac{\dots}{\dots} = \dots\dots\dots$ **(2 marks)**

2 Work out

> Convert all mixed numbers to improper fractions.

(a) $1\frac{1}{5} \times 2\frac{2}{3}$

$= \frac{\dots}{5} \times \frac{\dots}{3} = \frac{\dots}{\dots}$

(2 marks)

(b) $6\frac{7}{8} \div 3\frac{3}{4}$

$= \frac{\dots}{8} \div \frac{\dots}{4} = \frac{\dots}{8} \times \frac{\dots}{\dots}$

$= \frac{\dots}{\dots} = \dots\dots\dots$ **(2 marks)**

3 Work out

(a) $2\frac{3}{5} \times 4\frac{4}{9}$

$= \dots\dots \frac{\dots}{\dots}$ **(2 marks)**

(b) $9\frac{3}{8} \div 2\frac{13}{16}$

$= \dots\dots \frac{\dots}{\dots}$ **(2 marks)**

4 A shopping bag contains $5\frac{3}{4}$ kg of potatoes and another is filled with $3\frac{4}{7}$ kg of potatoes. What is the total weight of the potatoes?

$= 5\frac{\dots}{28} + 3\frac{\dots}{28} = \dots\dots \frac{\dots}{\dots} = \dots\dots \frac{\dots}{\dots}$ kg   **(3 marks)**

PROBLEM SOLVED!

5 A 195 km trip was completed by train, car and bicycle. $\frac{2}{5}$ of the distance was completed by car, $\frac{1}{3}$ by bicycle. How many kilometres was the train journey?

You will need to use problem-solving skills throughout your exam – **be prepared!**

.... km   **(3 marks)**

6 The distance between Portston and Chidhook is $18\frac{2}{5}$ miles. The distance between Brigville and Dimchester is $3\frac{1}{4}$ times that distance. What is the distance between Brigville and Dimchester?

You will need to use problem-solving skills throughout your exam – **be prepared!**

.... miles   **(3 marks)**

Calculator and number skills

1 Work out

> Order of operations (**BIDMAS**)
> **B**rackets, **I**ndices, **D**ivision, **M**ultiplication, **A**ddition, **S**ubtraction.

(a) $12 - 6 \div 2$

= $12 -$ .... = ... **(1 mark)**

(b) $17 - 3 \div 2 + 5$

= .... = ... **(1 mark)**

(c) $6 + (12 \times 3) \div 9$

= .... **(1 mark)**

(d) $\sqrt{(1 \div 6 \times 8)}$

> A root can be written as an index, so give it the same ranking in BIDMAS.

= $\sqrt{......}$ **(2 marks)**

2 Work out

(a) $\dfrac{23 + (17 + 2)^2 - 2}{6 + 5 \times 3}$

$= \dfrac{23 + 19^2 - 2}{6 + 15} = \dfrac{......}{21}$

= .... **(2 marks)**

(b) $\dfrac{3 + 8 + 2^3 + 11}{2 + \sqrt{9}}$

$= \dfrac{..........}{2 + 3} = \dfrac{......}{......}$

= .... **(2 marks)**

(c) $\dfrac{5 - 2}{5 \times 2 - 28 \div 7} + \dfrac{3^2 + -10 + 6}{\sqrt{25} + 10 \div 2}$

$= \dfrac{......}{......} = \dfrac{......}{......}$

= .... **(2 marks)**

3 Use a calculator to work out

$\dfrac{\sqrt{2.25} + 3.1}{3.1^2 - 2.5 \times 2.924}$

> Write down all the figures from your display and use BIDMAS.

$= \dfrac{1.5 + 3.1}{...... - ...... \times ......}$

= ....

> You can enter in this calculation all at once by using the fraction button. Calculators have BIDMAS built into their memory.
> $(S \Leftrightarrow D)$ This button converts between fractions and decimals. Give it a try.

(3 marks)

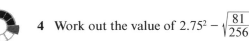

4 Work out the value of $2.75^2 - \sqrt{\dfrac{81}{256}}$

.... **(2 marks)**

5 Work out the value of $\dfrac{\sqrt[3]{31.255875} \times \sqrt{5.8564}}{2.35 \times \sqrt{2.4336}}$

(a) Write out all the digits displayed on the calculator.

.... **(2 marks)**

(b) Write this answer to 2 significant figures.

.... **(1 mark)**

Standard form 1

1 (a) Write 3500 in standard form.

$3500 = 3.5 \times 1000 = 3.5 \times 10^{\cdots}$ **(1 mark)**

(b) Write 1.21×10^{-2} as an ordinary number.

> The index is −2, therefore the figures will move two places to the right (making the number 100 times smaller).

....   **(1 mark)**

(c) Write 0.32×10^4 as an ordinary number.

....   **(1 mark)**

2 Write the following numbers in standard form.

(a) 600

.... **(1 mark)**

(b) 570

.... **(1 mark)**

(c) 4003

.... **(1 mark)**

(d) 51.1

.... **(1 mark)**

(e) 61.13

....   **(1 mark)**

3 Write the following in standard form.

(a) 0.35

$= 3.5 \times 10^{\cdots}$

> Rewrite each number so that it lies between 1 and 10. Now decide the power of 10 that it must be multiplied by to get it back to its original value.

    (1 mark)

(b) 0.00182

$0.00182 = 1.82 \times 10^{\cdots}$ **(1 mark)**

(c) 0.487×10^5

.... **(1 mark)**

4 (a) Write 5 842 000 in standard form.

$5842000 = 5.842 \times 10^{\cdots}$ **(1 mark)**

Use the following information $x = 5\ 842\ 000$, $y = 3.16 \times 10^3$.

(b) Work out $x + y$, giving your answer in standard form to 3 significant figures.

> Convert y into an ordinary number, work out $x + y$ and then convert the result to standard form.

....   **(2 marks)**

(c) Work out $x - y$, giving your answer in standard form to 3 significant figures.

....   **(2 marks)**

5 Voyager 2 took 4 years (approximately 3.5×10^4 hours) to get from Earth to Saturn. The distance between the two planets is approximately 7.46×10^8 miles. Work out the approximate speed of Voyager 2's journey to 3 significant figures.

> You will need to use problem-solving skills throughout your exam – **be prepared!**

> $speed = \dfrac{distance}{time}$

.... mph   **(4 marks)**

Standard form 2

6   Work out, giving your answer in standard form:

> Use the commutative property of multiplication which means you can multiply in any order.

(a) $(14 \times 10^3) \times (2 \times 10^{-1})$

$(14 \times 2) \times (10^3 \times 10^{-1}) = (14 \times 2) \times 10^{3-1} = (14 \times 2) \times 10^2$

....   **(2 marks)**

(b) $(12 \times 10^5) \div (24 \times 10^{-3})$ > Write it as a fraction, tidy up and cancel.

$\dfrac{12 \times 10^5}{24 \times 10^{-3}} = \dfrac{12}{24} \times \dfrac{10^5}{10^{-3}} = \dfrac{12}{24} \times 10^{5-(-3)}$

$= \dfrac{1}{2} \times 10^{\cdots}$

$= \ldots\ldots\ldots\ldots\ldots$ **(2 marks)**

7   Work out, giving your answer in standard form:

(a) $(3.2 \times 10^2) + (1.5 \times 10^3)$ (b) $(5.8 \times 10^6) - (7.3 \times 10^5)$

$\begin{array}{r} 320 \\ +1500 \\ \hline 1820 \end{array}$ $\begin{array}{r} 5\,800\,000 \\ -\ \ 730\,000 \\ \hline 5\,070\,000 \end{array}$

$= 1.82 \times 10^{\cdots}$ **(2 marks)**

.... **(2 marks)**

8   If $P = (3.2 \times 10^3)$ and $Q = (7 \times 10^{-3})$ solve the following, leaving your answer in standard form.

(a) $2P$ (b) $P \times Q$

$= 2 \times (3.2 \times 10^3) = \ldots\ldots$ $= (3.2 \times 10^3) \times (7 \times 10^{-3}) = \ldots\ldots$

                     (2 marks)   **(2 marks)**

(c) $P - 2Q$ (d) $3P \div 2Q$

$= (3.2 \times 10^3) - 2 \times (7 \times 10^{-3}) = \ldots\ldots$ ....

                     (2 marks)   **(2 marks)**

9   The surface area of the earth is approximately 510 million km²; the surface area of the moon is approximately 3.8×10^7 km². Which has the larger surface area and how many times greater is the larger surface area?
Write your answer to 3 significant figures.

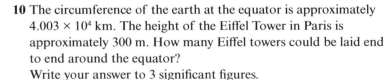

....   **(3 marks)**

10 The circumference of the earth at the equator is approximately 4.003×10^4 km. The height of the Eiffel Tower in Paris is approximately 300 m. How many Eiffel towers could be laid end to end around the equator?
Write your answer to 3 significant figures.

....   **(3 marks)**

Counting strategies

1 Majid has three letters printed on cards P, Q and R. He also has another three cards with the numbers 2, 4 and 6 printed. Find all the ways he can pair a letter with a number.

Ⓟ ②
Ⓠ ④
Ⓡ ⑥

> Remember that it does not matter which order they appear.

(P, 2), (P, …), (…, 6), (Q, …), (…, …), (…, …), (…, …), (…, …), (…, …)   **(2 marks)**

2 Millie wants to buy a new bicycle. She can choose a touring bike, a mountain bike or a fold-up bike. Her colour choices are orange, jade or silver. List all her possible choices of bicycle.

> Use the first letter of the bicycle and the colour choices.

	Touring (T)	Mountain (M)	Fold-up (F)
Orange (O)			
Jade (J)			
Silver (S)			

.... **(2 marks)**

3 There are 4 coins on the table: 10p, 20p, 50p and £1. List all the different sums that can be made with any 2 different coins.

	10p	20p	50p	£1
£1				
50p				
20p				
10p				

= ....   **(2 marks)**

4 Three cards each have a different digit on them, 4, 5 and 6. Write down all the six numbers you can make with these cards.

> Start with the first card: how many ways can that be the first digit?

.... **(2 marks)**

5 Five musical instruments, guitar, piano, trumpet, flute and mandolin, are to be recorded on a CD but there will only be two instruments on each track. Each instrument will be recorded with each of the others. How many tracks will be on the CD?

> You will need to use problem-solving skills throughout your exam – **be prepared!**

....   **(3 marks)**

Had a go ☐ **Nearly there** ☐ **Nailed it!** ☐

Problem-solving practice 1

1 It is said that every even number is the sum of two prime numbers. Find two prime numbers that sum to the following even numbers.

(a) 8

.... + ... **(1 mark)**

(b) 12

.... + ... **(1 mark)**

(c) 20

.... + ... **(1 mark)**

(d) 36

.... + ... **(1 mark)**

2 16 GB flash drives cost £5.99 and 32 GB flash drives cost £6.88. How much change would you get from £50 if you bought three 16 GB and four 32 GB flash drives?

.... **(3 marks)**

3 A hot chocolate costs £2.70. A group of friends give the barista £25 to pay for their hot chocolates. What is the maximum number of people there could be in the group, and how much change would there be for that number of hot chocolates?

.... **(2 marks)**

4 A hotel buys packets of biscuits which each contain 14 biscuits. The hotel is running a conference and plans to provide each person at the conference with 3 biscuits during their coffee break. There are 800 people booked for the conference. How many packets of biscuits will the hotel need to buy?

.... **(3 marks)**

5 Which fraction is larger: $\frac{3}{5}$ or $\frac{5}{7}$? Show your working.

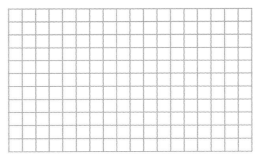

$\frac{3}{5} = \frac{3 \times \ldots\ldots}{5 \times \ldots\ldots}$ $\frac{5}{7} = \frac{5 \times \ldots\ldots}{7 \times \ldots\ldots}$

.... **(2 marks)**

6 Ice cream cornets are produced at a rate of 200 every hour. The factory producing them operates for $6\frac{1}{2}$ hours per day, 5 days each week. There are 8 ice cream cornets packed in a box. How many boxes are needed each week?

.... **(4 marks)**

Problem-solving practice 2

7 In a triangle, angle $A = \frac{1}{2}$ of the total of all the angles, and angle $B = \frac{1}{6}$ of the total of all the angles.

(a) What fraction of all angles in the triangle is angle C?

.... **(2 marks)**

(b) What is the size of angle C?

.... **(2 marks)**

8 Misha buys an 8 kg bag of flour to make pizzas. Each pizza uses 300 g of flour. She sells the pizzas in her restaurant for £8.50 each.

(a) How many pizzas can she make?

.... **(3 marks)**

(b) How much money does she make if she sells them all?

.... **(2 marks)**

9 Deepak has three planks of wood left over from a maintenance job in his house. One piece is 85 cm long, one is 79 cm and the other is 83 cm long. He now needs to cut 7 pieces from what is left, all of the same length for the last part of the job. What is the maximum size for each piece?

.... **(4 marks)**

10 Two cyclists, Anne and Margaret, are riding on a velodrome. They start together. It takes Anne 30 seconds to complete one circuit. It takes Margaret 40 seconds to complete a circuit. How many circuits will they each have completed when they are both together at the starting point again?

Anne ....

Margaret....................... **(3 marks)**

11 A train is travelling at a constant velocity of 112 km/h. Write this velocity as m/s (metres per second).

.... **(5 marks)**

12 What is the value of n in the following equation?

$$5^n \times 5^{3n} = \frac{5^5 \times 5^4}{5^3}$$

.... **(4 marks)**

21

Collecting like terms

1 Choose from term, **expression**, **equation** and **formula** to describe the following.

> Term: In $2t + r^2$ there are two terms, $2t$ and r^2.
> Expression: May contain numbers, variables (letters) and operators.
> Equation: Contains an = sign.
> Formula: A mathematical rule described using symbols.

(a) $3x - 4 = 13$

.... **(1 mark)**

(b) $5t$

.... **(1 mark)**

(c) $y = ax + b$

.... **(1 mark)**

(d) $p^2 - 2p = 15$

.... **(1 mark)**

(e) $s = \dfrac{d}{t}$

.... **(1 mark)**

2 Simplify the following.

(a) $n + 3n + n + n - 2$

....

> Like terms contain the same variable, or multiplied group of variables. One example of a pair of like terms is xy and $3xy$; another is $2x^2$ and $\frac{1}{2}x^2$.

(1 mark)

(b) $3ab + 6ab - ab$

....

> Add up all the like terms.
> For example: $2abc + 3abc + abc = (2 + 3 + 1)abc = 6abc$

(1 mark)

3 Simplify

(a) $p + 2q + 3p + q + 4p$

.........p + ...p + ...p + ...q + ...q = ...p + ...q **(2 marks)**

(b) $15pqr - 6pqr + pqr$

.... pqr **(2 marks)**

4 Simplify

(a) $15x - 8x$

.... **(1 mark)**

(b) $2y^2 + 8y^2$

.... **(1 mark)**

(c) $6x^2 + 3x + 2x^2 - x$

.... **(2 marks)**

(d) $7 + 2n + 23 - 5n$

.... **(2 marks)**

5 Simplify

(a) $a^2 + a^2 + a^2$

$= \ldots\ldots a^2$ **(1 mark)**

(b) $a^2 + a + 2a^2 - 3a + b$

$= \ldots\ldots a^2 + \ldots\ldots a + \ldots\ldots b$ **(2 marks)**

(c) $x + 2x^2 + 5 - 3x + x^2$

.... **(2 marks)**

Simplifying expressions

 1 Simplify

> Remember $x + x = 2x$ and $2x \times 3x = 6x^2$.

(a) $a \times a \times a$

$a^{......}$ **(1 mark)**

(b) $7a \times b$

$7 \times a \times b = \ldots\ldots\ldots\ldots$ **(1 mark)**

(c) $p \times 2q \times r$

$p \times 2 \times q \times r = \ldots\ldots\ldots\ldots\ldots\ldots\ldots\ldots\ldots$

 2 Simplify

> Use the commutative property of multiplication (you can do it in any order) and rearrange so that the numbers are multiplied first.

(a) $b \times b \times 2b$

$= 2b^{...}$ **(1 mark)**

(b) $3 \times a \times 3 \times b$

$= 9 \ldots\ldots\ldots\ldots\ldots$ **(1 mark)**

(c) $10q \times 2$

$\ldots\ldots\ldots\ldots\ldots\ldots$ **(1 mark)**

(d) $11g \times 2h$

$\ldots\ldots\ldots\ldots\ldots\ldots$ **(1 mark)**

 3 Simplify

(a) $3y \times 2y$

$= 3 \times y \times 2 \times y = 3 \times 2 \times y \times y$

$= \ldots\ldots\ldots\ldots\ldots\ldots y^{.....}$ **(1 mark)**

(b) $6n \times 3m$

$= 6 \times 3 \times m \times n$

$= \ldots\ldots\ldots\ldots\ldots\ldots$ **(1 mark)**

(c) $15a \div 5$

$= \dfrac{15a}{5}$

$= \ldots\ldots\ldots\ldots\ldots\ldots$ **(1 mark)**

(d) $30xy \div 2y$

$= \dfrac{30 \times x \times y}{2 \times y}$

$= \ldots\ldots\ldots\ldots\ldots\ldots$ **(1 mark)**

 4 Simplify

 Guided

(a) $7a \times 5c$

$\ldots\ldots\ldots\ldots\ldots\ldots$ **(1 mark)**

(b) $3u \times 2u \times 5u$

$\ldots\ldots\ldots\ldots\ldots\ldots$ **(1 mark)**

(c) $48fg \div 8f$

$\ldots\ldots\ldots\ldots\ldots\ldots$ **(1 mark)**

(d) $27abc \div 9bc$

$\ldots\ldots\ldots\ldots\ldots\ldots$ **(1 mark)**

 5 Simplify

  Guided

(a) $q \times 4p \times 5r$

$\ldots\ldots\ldots\ldots\ldots\ldots$ **(1 mark)**

(b) $72efg \div 12eg$

$\ldots\ldots\ldots\ldots\ldots\ldots$ **(1 mark)**

 6 Which expression below is **not** equal to any of the others?

 Guided

$8xy^2$ $\dfrac{12x^2y^2}{4y}$ $\dfrac{16x^2y^2}{2x}$ $2x \times 2x$ $3x^2y$

$\ldots\ldots\ldots\ldots\ldots\ldots$ **(2 marks)**

Algebraic indices

1 Simplify and leave your answers in index form.

> When bases are the same, add the indices. Here the base is y.

(a) $y^2 \times y^4 = y^{2+\cdots\cdots} = \ldots\ldots\ldots\ldots$ **(1 mark)**

> When bases are the same, **subtract** the index in the denominator from that in the numerator. Here the base is n.

(b) $n^5 \div n^3 = \dfrac{n^5}{n^3} = n^{5-\cdots\cdots} = \ldots\ldots\ldots\ldots$ **(1 mark)**

> Multiply the index inside the bracket by the one on the outside.

(c) $(f^2)^3 = f^2 \times f^2 \times f^2 = f^{2\times\cdots\cdots} = \ldots\ldots\ldots\ldots$ **(1 mark)**

2 Simplify and leave your answers in index form.

(a) $\dfrac{a^5 \times a^3}{a^2}$

.... **(2 marks)**

(b) $\dfrac{q^{15}}{q^3 \times q^5}$

.... **(1 mark)**

(c) $\dfrac{b^4 \times b^3}{b^2 \times b}$

.... **(2 marks)**

(d) $\dfrac{c^9 \times c^{-2}}{c^5 \times c^{-6}}$

.... **(1 mark)**

3 Simplify

(a) $(b^4)^5$

$= b^4 \times b^4 \times b^4 \times b^4 \times b^4$

.... **(1 mark)**

(b) $(2y^2)^3$

$= 2y^2 \times 2y^2 \times 2y^2$

.... **(1 mark)**

(c) $(3a^3)^2 \times 2a$

.... **(2 marks)**

> Any variable (letter) without an index can be written raised to the power 1. For example, a can be written as a^1.

(d) $5x^2 \times 2x^5$

.... **(1 mark)**

(e) $12a^4 \div 2a^3b$. Write your answer as a fraction.

$\dfrac{12a^4}{2a^3b} = \dfrac{12 \times a \times a \times a \times a}{2 \times a \times a \times a \times b}$

.... **(2 marks)**

(f) $8x^2y^5 \div 2xy^2$

.... **(2 marks)**

PROBLEM SOLVED!

4 Work out the value of n.

$x^2 \times x^{5n} = \dfrac{x^9 \times x^7}{x^4}$

> You will need to use problem-solving skills throughout your exam – **be prepared!**

$n = \ldots\ldots\ldots\ldots\ldots\ldots\ldots\ldots\ldots\ldots$ **(2 marks)**

Substitution

1 Use the formula speed = $\frac{\text{distance}}{\text{time}}$ to calculate the time it takes a car to travel a distance of 80 miles at 50 mph. Write your answer as a decimal.

Time = $\frac{\dotsb}{\dotsb}$ = ... hours **(1 mark)**

2 Work out the value of

> You can apply brackets to keep track of your calculations and to separate variables (letters) from constants (numbers).

(a) $7x + 11$ when $x = 5$

$(7 \times \dots) + 11 = \dots$ **(2 marks)**

(b) $3x - 12$ when $x = 7$

$(3 \times \dots) - 12 = \dots$ **(2 marks)**

3 Work out the value of

(a) $12n + 3m$ when $n = 2, m = 3$

.... **(2 marks)**

(b) $7a - 3b$ when $a = 4, b = -3$

.... **(2 marks)**

4 Work out the value of

(a) $6p + qr$ when $p = 3, r = -4$

> Take care when multiplying negative numbers
> $- \times - = +$ $\quad - \times + = -$ $\quad + \times - = -$

$(6 \times \dots) + (\dots \times \dots) = \dots$ **(2 marks)**

(b) $5a^3$ when $a = 2$

$5 \times (\dots \times \dots \times \dots) = \dots$

(2 marks)

(c) $4r^2 - 3q$ when $r = 3, q = -3$

$4 \times (\dots)^2 - 3 \times (\dots) = \dots$

(2 marks)

5 Work out the value of

(a) $x(3 - 2y)$ when $x = 3, y = -3$

.... **(2 marks)**

(b) $4p - q(5 + r)$ when $p = 5, q = -1, r = 3$

.... **(2 marks)**

(c) $\frac{ab^2}{a - b}$ when $a = 4, b = 2$

> ab^2 is not the same as $(ab)^2$
> ab^2 means $a \times b \times b$
> $(ab)^2$ means $(a \times b) \times (a \times b)$

.... **(2 marks)**

6 The volume of a sphere is found by using the formula $\frac{4}{3}\pi r^3$. If we say that π is approximately 3 and r = radius, what would the radius be if the volume of a sphere was given as 32 cm³?

$\frac{4 \times \pi \times r \times r \times r}{3} = 32$

> Write the formula out as a fraction and turn the information given into an equation.

> You will need to use problem-solving skills throughout your exam – **be prepared!**

So $r^3 = \frac{\dots}{\dots} = \dots$

$r = \dots$ cm **(2 marks)**

Formulae

1 The German mathematician Euler proved that any polyhedron (a 3D object made of 2D polygons) can be described by the formula $V - E + F = 2$ where V = vertices, E = edges and F = faces. Work out how many faces a polyhedron would have if it had 8 vertices and 12 edges.

$V - E + F = 2$

So, $8 - 12 + F = 2$

$F = \dots\dots\dots\dots\dots$ **(2 marks)**

2 An electricity company uses the formula $C = 0.21k + 35$ to calculate a monthly bill for its users in £s. In the formula, 35 is the standing monthly charge (£), C is the cost and k stands for the number of kilowatt hours of electricity used. If Jodie had a bill for £161.40, how much electricity (kwh) did she consume in that month? Give your answer to 3 significant figures.

> Form an equation and substitute in the value of 161.40 for C.

$\dots\dots\dots\dots\dots$ kw **(2 marks)**

3 The height of a ladder in cm is given by the formula $h = 35r + 40$ where r is the number of rungs on the ladder.

(a) Work out the height of a ladder with 18 rungs. Give your answer in metres.

$\dots\dots\dots\dots\dots$ m **(2 marks)**

(b) How would you explain in words what 35 represents in the formula?

> Drawing a diagram of the first three rungs of the ladder might be helpful to visualise it.

$\dots\dots\dots\dots\dots\dots\dots\dots\dots\dots\dots\dots\dots\dots\dots\dots\dots\dots\dots\dots\dots\dots\dots\dots\dots$ **(1 mark)**

4 George was told that his BMI (Body Mass Index) was 31.2. He looked at the formula for this:

$\text{BMI} = \dfrac{m}{h^2}$

where m = mass in kg and h = height in metres. His height is 1.75 m. What is his mass? Give your answer to the nearest kg.

> To work out the mass, m, rearrange the equation to make m the subject and then insert the values for BMI and h.

$\dots\dots\dots\dots\dots$ **(2 marks)**

5 The distance an object drops is found from the formula $d = \dfrac{g}{2} \times t^2$ where d = distance (metres), g = gravity and t = time (s). If we take the value of g to be 10, how far would an object have fallen if it took 8 seconds to fall?

$\dots\dots\dots\dots\dots$ **(2 marks)**

6 The formula from question 5 can be rearranged to $t = \sqrt{\dfrac{2d}{g}}$ to calculate the time in seconds that an object takes to fall to the ground. Using the same value of 10 for gravity and a distance (height) of 90 m, how long would the object take to hit the ground?

> Work out the number inside the square root first, then take the positive square root, since time taken cannot be negative.

$\dots\dots\dots\dots\dots$ **(2 marks)**

7 A formula states $f = 2g^2 - 5g + 3$. Show that if $f = 21$ then $g = -2$.

$\dots\dots\dots\dots\dots$ **(2 marks)**

Writing formulae

1 Packets of steel guitar strings cost £*M* and packets of nylon guitar strings cost £*N*.

(a) Write a formula in terms of *M* and *N* for the cost (*C*) of 2 packets of steel strings and 5 packets of nylon strings.

$C = (\ldots\ldots) \times M + (\ldots\ldots) \times N = (\ldots\ldots)M + (\ldots\ldots)N$

....

(2 marks)

(b) If *M* = £8.50 and *N* = £7.45, use the formula to work out the cost of 7 packets of steel strings plus 2 packets of nylon strings.

....

(1 mark)

2 Steel guitar strings (*M*) cost £2 more than nylon guitar strings (*N*). Write down the formula for the cost of *M* in terms of *N*.

Cost of nylon strings = £ ....

Cost of steel strings equals cost of nylon + £2

....

> You will need to use problem-solving skills throughout your exam – **be prepared!**

(2 marks)

3 A cuboid is labelled in terms of *r* and *q*.

(a) Write a formula for the volume (*V*) of the cuboid. Use index form in your answer.

$V = r \times q \times 2r = \ldots\ldots\ldots$

....

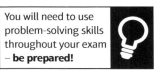

(2 marks)

(b) Write a formula for the total surface area (*A*) of the cuboid. Use index form where necessary.

> Work out the area of each of the 6 sides of the cuboid and then add them all together. (Notice that there are three pairs of the same area.)

....

(2 marks)

4 The total age of Andy (*A*) plus Bahti (*B*) plus Cathy (*C*) is 20 years.

(a) Write down a formula for the sum of their ages.

> You will need to use problem-solving skills throughout your exam – **be prepared!**

....

(2 marks)

(b) If Andy's age is equal to the sum of Bahti's and Cathy's ages and Bahti is 2 years younger than Cathy

(i) write the formula for *A* in terms of *B* and *C*

....

(2 marks)

(ii) write a formula for *A* in terms of *B*.

....

(2 marks)

Expanding brackets

1 Expand the following.

(a) $5(x + 1)$

(5 × ....) + (5 × ....)

.... **(2 marks)**

(b) $3(2x - 1)$

(3 × 2x) + (3 × ....)

.... **(2 marks)**

(c) $\sqrt{3}(x + \sqrt{3})$ $\boxed{\sqrt{n} \times \sqrt{n} = n}$

....   **(2 marks)**

2 Expand the following.

(a) $-2(x + 3)$

(−2 × ....) + (−2 × ....)

= .... + ... **(2 marks)**

(c) $-3(4 - 2x)$

(3 × 2x) + (3 × ....)

.... **(2 marks)**

(b) $-4(2x + 1)$

.... **(2 marks)**

> If there is a negative sign by itself in front of the brackets you multiply both terms in the brackets by −1.

3 Expand the following.

(a) $x(x + 3)$

(x × ....) + (x × ....)

.... **(2 marks)**

(b) $2x(2x - 5)$

(2x × ....) + (2x × [− ....])

.... **(2 marks)**

(c) $-5x(-3y - 2x)$

.... **(2 marks)**

4 Expand the following.

(a) $2x + 4(x + 3)$

2x + (4 × ....) + (4 × ....)

.... **(2 marks)**

(b) $2(x - 1) + 3(x - 2)$

(2 × ....) + (2 × − ....) + (3 × ....) + (3 × − ....)

.... **(2 marks)**

5 $2x(x + 3) - x(5x - 2)$ can be written as an expression of the form $ax^2 + bx$ where a and b are whole positive or negative numbers. Work out the values of a and b in the expression.

 Guided

a = ... **(2 marks)**

b = ... **(2 marks)**

6 A line A is measured and recorded as $3x + 2$ cm. A line B is measured and recorded as $2x - 1$ cm. What would be the total length of a line measured and recorded as 3A + 2B?

 Guided

 PROBLEM SOLVED!

> You will need to use problem-solving skills throughout your exam – **be prepared!**

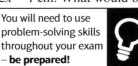

> Write 3A + 2B in terms of x.

3A = ... **(2 marks)**

2B = ... **(2 marks)**

3A + 2B = ... **(1 mark)**

Factorising

1 Factorise the following.

(a) $2x + 8$

$2(\ldots\ldots\ldots + \ldots\ldots\ldots)$ **(1 mark)**

> Look for the HCF of the number parts of the expressions.
> For example: $10x - 20$
> HCF (10, 20) = 10, so $10x - 20 = 10(x - 2)$

(b) $12x - 36$

$12(\ldots\ldots\ldots - \ldots\ldots\ldots)$ **(1 mark)**

(c) $6p - 30$

$\ldots\ldots\ldots\ldots\ldots\ldots\ldots\ldots\ldots\ldots\ldots\ldots$ **(1 mark)**

2 Factorise the following.

(a) $x^2 + 5x$

$x(x + \ldots\ldots\ldots)$ **(2 marks)**

> You need to look for the HCF of the lettered part of these expressions.
> HCF of x^2 and x is x.

(b) $x^2 - 14x$

$x(\ldots\ldots\ldots - \ldots\ldots\ldots)$ **(2 marks)**

(c) $a^2 + 17a$

$\ldots\ldots\ldots\ldots\ldots\ldots\ldots\ldots\ldots\ldots\ldots\ldots$ **(2 marks)**

3 Fully factorise the following.

(a) $2x^2 + 4x$

> Fully factorise means you need to find the HCF of both terms in the expression then take it outside of brackets.

$2x(\ldots\ldots\ldots + \ldots\ldots\ldots)$ **(2 marks)**

(b) $27r^2 - 9r$

$\ldots\ldots\ldots\ldots\ldots\ldots\ldots\ldots\ldots\ldots\ldots\ldots$ **(2 marks)**

(c) $8t^2 - 64t$

$\ldots\ldots\ldots\ldots\ldots\ldots\ldots\ldots\ldots\ldots\ldots\ldots$ **(2 marks)**

4 Fully factorise

(a) $8e^2 + 12e$

$= 4e(2e + \ldots\ldots\ldots)$

> $2e(4e - 6)$ is not fully factorised.

(1 mark)

(b) $12x^2 - 16x$

$\ldots\ldots\ldots\ldots\ldots\ldots\ldots\ldots\ldots\ldots\ldots\ldots$ **(2 marks)**

(c) $14q^2 - 35q$

$\ldots\ldots\ldots\ldots\ldots\ldots\ldots\ldots\ldots\ldots\ldots\ldots$ **(2 marks)**

5 Here is a rectangle with an area of $3x^2 + 12x$. Fully factorise $3x^2 + 12x$ to find its length (L) and width (W).

$3x^2 + 12x$

L = $\ldots\ldots\ldots\ldots\ldots\ldots\ldots$ **(2 marks)**

W = $\ldots\ldots\ldots\ldots\ldots\ldots\ldots$ **(2 marks)**

6 Why is the value of $3x^2 + 5xy$ always an even number if x and y are both odd numbers?

$\ldots\ldots\ldots\ldots\ldots\ldots\ldots\ldots\ldots\ldots\ldots\ldots\ldots\ldots\ldots\ldots\ldots\ldots\ldots\ldots\ldots\ldots\ldots\ldots\ldots\ldots$

$\ldots\ldots\ldots\ldots\ldots\ldots\ldots\ldots\ldots\ldots\ldots\ldots\ldots\ldots\ldots\ldots\ldots\ldots\ldots\ldots\ldots\ldots\ldots\ldots\ldots\ldots$

$\ldots\ldots\ldots\ldots\ldots\ldots\ldots\ldots\ldots\ldots\ldots\ldots\ldots\ldots\ldots\ldots\ldots\ldots\ldots\ldots\ldots\ldots\ldots\ldots\ldots\ldots$

$\ldots\ldots\ldots\ldots\ldots\ldots\ldots\ldots\ldots\ldots\ldots\ldots\ldots\ldots\ldots\ldots\ldots\ldots\ldots\ldots\ldots\ldots\ldots\ldots\ldots\ldots$

> You will need to use problem-solving skills throughout your exam
> – **be prepared!**

PROBLEM SOLVED!

(3 marks)

Linear equations 1

 1 Solve the following linear equations.

(a) $3x = 18$

$3x = 18 \Rightarrow \dfrac{3x}{3} = \dfrac{18}{3}$

.... **(1 mark)**

> A letter is referred to as an unknown. If you want to remove a number which is multiplying an unknown, you should divide by that number.

(b) $7a = -35$

$\dfrac{7a}{7} = \dfrac{-35}{7}$

So $a =$ .... **(1 mark)**

(c) $\dfrac{r}{-4} = 5$

.... **(1 mark)**

> If an unknown is being divided by a number, to leave the unknown by itself you must multiply both sides of the equation by that number to keep the equation balanced. For example, $\frac{x}{5} = 3$ becomes $\frac{5 \times x}{5} = 5 \times 3$. Cancelling the 5s on the left gives $x = 15$.

> When multiplying a fraction by any integer, only multiply the numerator.

 2 Find the value of the unknown.

(a) $x - 3 = 11$

$x - 3 + 3 = 11 + 3$

.... **(2 marks)**

(b) $q + 11 = 16$

$q + 11 - 11 = 16 - 11$

.... **(2 marks)**

 3 Solve the following.

(a) $2x - 9 = 5$

$2x - 9 + 9 = 5 + 9$

$\dfrac{2x}{2} = \dfrac{......}{2}$

$x =$ .... **(2 marks)**

(b) $5q + 23 = 7$

$5q = 7 - 23$

$5q = -16$

$q = \dfrac{......}{......}$ **(2 marks)**

(c) $6a + 13 = 7$

.... **(2 marks)**

 4 The diagram shows an irregular quadrilateral.

> You will need to use problem-solving skills throughout your exam – **be prepared!**

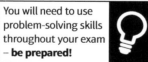 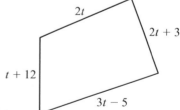

(a) Write an equation for the perimeter (P) in terms of t.

$P =$ .... **(2 marks)**

(b) If $t = 3\,$cm, find the value of P

$P =$ .... **(1 mark)**

(c) If $P = 82$ cm, find the value of t

$t =$ .... **(2 marks)**

Linear equations 2

5 Solve the following linear equations.

(a) $2x + 5 = 5x - 4$

$2x - 2x + 5 = 5x - 2x - 4$

$5 = 3x - 4$

$5 + 4 = 3x - 4 + 4$

$......... = 3x$

$x = ...$

> Aim to get unknowns on one side of the equals sign and constants on the other using addition or subtraction first, followed by multiplication or division.

(2 marks)

(b) $2a - 3 = a + 5$

$2a - a - 3 = a - a + 5$

$a - 3 + 3 = 5 + 3$

$a = ...$

(2 marks)

(c) $5q - 7 = 3q - 1$

$5q - 3q - 7 = 3q - 3q - 1$

$2q ....$

$q = ...$

(2 marks)

6 Find the value of the unknown.

(a) $4(3x - 4) = 20$

$3x - 4 = \dfrac{20}{4} = 5$

$3x = ...$

$x = ...$

> First divide both sides by any common factor (if there is one). Then multiply out the brackets and collect like terms.

(2 marks)

(b) $3(3t - 2) = 5t + 2$

$t = ...$

> To rewrite an equation without a fraction term, multiply both sides by its denominator.

(2 marks)

(c) $\dfrac{5 - 2u}{3} = 1$

$u = ...$

(2 marks)

7 Both shapes shown below have the same area. Work out the value of n.

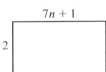
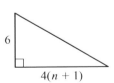

> The area of a triangle is found by multiplying the base by the vertical height, then dividing the result by 2. The area of a rectangle is found by multiplying base by height. Use these facts to write an equation in n.

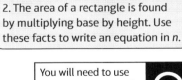

> You will need to use problem-solving skills throughout your exam – **be prepared!**

$n = ...$

(3 marks)

Inequalities

1 Write down the inequalities shown on the number line.

(a)

$$-5\,-4\,-3\,-2\,-1\ 0\ 1\ 2\ 3\ 4\ 5$$

> Closed circles (filled in) show numbers that are included. Open circles show numbers that are not included.

$x \geqslant$ .... **(1 mark)**

(b)

$$-5\,-4\,-3\,-2\,-1\ 0\ 1\ 2\ 3\ 4\ 5$$

$x >$ .... **(1 mark)**

(c)

$$-5\,-4\,-3\,-2\,-1\ 0\ 1\ 2\ 3\ 4\ 5$$

.... $< x \leqslant$ .... **(1 mark)**

2 Show the inequality on the number line.

(a) $x \leqslant 2$

$$-5\,-4\,-3\,-2\,-1\ 0\ 1\ 2\ 3\ 4\ 5$$

(1 mark)

(b) $x > 2$

$$-5\,-4\,-3\,-2\,-1\ 0\ 1\ 2\ 3\ 4\ 5$$

(1 mark)

(c) $-1 \leqslant x \leqslant 2$

$$-5\,-4\,-3\,-2\,-1\ 0\ 1\ 2\ 3\ 4\ 5$$

(1 mark)

Guided

3 Write all the possible integers in the following inequalities.

> An integer is any whole number, negative, positive or zero.

(a) $-2 < x \leqslant 3$

$x = -1, 0,$ ...., ..., ..., **(1 mark)**

(b) $13 \leqslant x < 14$

$x =$ .... **(1 mark)**

(c) $-3.4 \leqslant x < 2.7$

$x =$ ...., ..., ..., 0, 1, ... **(1 mark)**

4 The numbers given have been rounded. Write an inequality that shows the range of actual values.

(a) $q = 40$ (to the nearest 10)

$35 \leqslant q < 45$ **(1 mark)**

(b) $r = 300$ (to the nearest 100)

$250 \leqslant r$ .... **(1 mark)**

(c) $u = 3.5$ (to 1 decimal place)

.... $u < 3.55$ **(1 mark)**

5 A marathon race was completed by Anna in 5 hours and 12 minutes to the nearest minute. Write an inequality that shows all possible times that she could have taken to complete the race.

> You will need to use problem-solving skills throughout your exam – **be prepared!**

Guided

PROBLEM SOLVED!

.... **(2 marks)**

Solving inequalities

1 Solve the following inequalities.

> Inequalities can for the most part be treated in the same way as equations.

(a) $3x \leqslant 15$

$$\frac{3x}{3} \leqslant \frac{15}{3}$$

$x > \dots\dots\dots$ **(1 mark)**

(b) $4x > 12$

> Check your result by substituting into the inequality two values for x that are greater and smaller than the value you have found.

$x > \dots\dots\dots$ **(1 mark)**

(c) $12x \leqslant 6$

$x \leqslant \dots\dots\dots$ **(1 mark)**

2 Solve

(a) $2x + 3 \leqslant 37$

> Subtract 3 from both sides.

$2x + 3 - 3 \leqslant 37 - 3$

$2x \leqslant 34$

$$\frac{2x}{2} \leqslant \frac{34}{2}$$

> Now divide both sides by 2.

$x \leqslant \dots\dots\dots$ **(2 marks)**

(b) $3x - 8 > 13$

$x > \dots\dots\dots$ **(2 marks)**

(c) $12x - 17 \leqslant 31$

$x \leqslant \dots\dots\dots$ **(2 marks)**

3 If x is an integer, what are all its possible values?

(a) $-10 \leqslant 2x < 4$

$$\frac{-10}{2} \leqslant \frac{2x}{2} < \frac{4}{2}$$

So $-5 \leqslant x < \dots\dots\dots$

$x = -5, -4, \dots\dots, \dots\dots, \dots\dots, \dots\dots \quad \dots\dots$ **(2 marks)**

(b) $1 \leqslant 4x \leqslant 14$

$x = \dots\dots\dots\dots\dots\dots\dots\dots$ **(2 marks)**

(c) $0 < 5x - 2 < 13$

$\dots\dots\dots\dots\dots\dots\dots\dots$ **(2 marks)**

4 Find the integer value of x that satisfies both equations $3x + 1 > 8$ and $5x - 13 < 12$.

> Solve each inequality in turn then compare.

$\dots\dots\dots\dots\dots\dots\dots\dots$ **(3 marks)**

Had a go ☐   **Nearly there** ☐   **Nailed it!** ☐

Sequences 1

1 Here is a pattern. Draw the next pattern in the sequence.

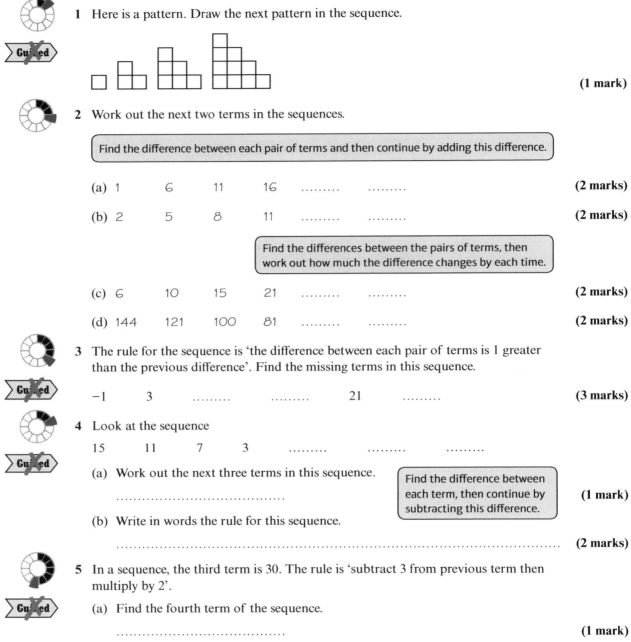

(1 mark)

2 Work out the next two terms in the sequences.

> Find the difference between each pair of terms and then continue by adding this difference.

(a) 1   6   11   16   ....   **(2 marks)**

(b) 2   5   8   11   ....   **(2 marks)**

> Find the differences between the pairs of terms, then work out how much the difference changes by each time.

(c) 6   10   15   21   ....   **(2 marks)**

(d) 144   121   100   81   ....   **(2 marks)**

3 The rule for the sequence is 'the difference between each pair of terms is 1 greater than the previous difference'. Find the missing terms in this sequence.

    −1   3   21   ....   **(3 marks)**

4 Look at the sequence

    15   11   7   3   ....

(a) Work out the next three terms in this sequence.

> Find the difference between each term, then continue by subtracting this difference.

    ....   **(1 mark)**

(b) Write in words the rule for this sequence.

    ....   **(2 marks)**

5 In a sequence, the third term is 30. The rule is 'subtract 3 from previous term then multiply by 2'.

(a) Find the fourth term of the sequence.

    ....   **(1 mark)**

(b) Find the first two terms of the sequence.

> Use the inverse of the rule to find the first two terms.

    ....   **(3 marks)**

ALGEBRA

Sequences 2

6 Write the general term (*n*th term) for the following sequences.

(a) 2   7   12   17

+5 +5 +5

2,   7,   12,   17

*n*th term = 5*n* − ....

> Remember the first *n* value is 1, the second is 2 and so on. What rule turns 1 into 2, 2 into 7, 3 into 12,...? The sequence goes up in 5s, so try multiplying each *n* value by 5 and finding what you need to take away.

(2 marks)

(b) 3   7   11   15

*n*th term = 4*n* − ....

(2 marks)

(c) 5   9   13   17

*n*th term = 4*n* + ...

(2 marks)

(d) 11   14   17   20

*n*th term = ...

(2 marks)

7 The first 4 terms of a sequence are given. Work out the *n*th term.

8   17   26   35

....

(2 marks)

8 Here is part of a number sequence. The first and fifth terms are missing.

....   3   15   27   ....

(a) Work out the 2 missing terms.

> Look for a pattern in the differences between the terms given.

....

(2 marks)

(b) Work out an expression for the *n*th term of the sequence.

> Find a multiplier for *n*, then work out what you need to take away.

....

(2 marks)

9 Look at the sequence

−4   −1   2   5   8

(a) Work out the *n*th term.

....

(2 marks)

(b) What is the 15th term of the sequence?

....

> Use your answer to part (a) to help you.

(2 marks)

35

Had a go ☐ **Nearly there** ☐ **Nailed it!** ☐

Coordinates

1 Two points, *A* and *B*, are marked on the grid.

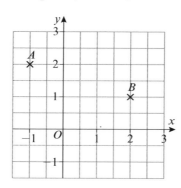

Coordinates are written (*x, y*). The first coordinate written is the distance from 0 in the *x* direction. The second is the distance from 0 in the *y* direction.

(a) Write down the coordinates of

    (i) point *B*

      (2, ...)   **(1 mark)**

    (ii) point *A*

      (..........., 2)   **(1 mark)**

(b) Mark the following points on the graph.

    (i) (−1, 1), label it *C*

      (1 mark)

    (ii) (2, −1), label it *D*

      (1 mark)

2 A line is drawn between the following pairs of coordinate points. Find the coordinates of the midpoints.

The midpoint between two points (a, b) and (c, d) has the coordinates $\left(\dfrac{a+c}{2}, \dfrac{b+d}{2}\right)$.

(a) (5, 8) and (13, 14)

$$\text{midpoint} = \left(\frac{5+13}{2}, \frac{\ldots\ldots + \ldots\ldots}{2}\right)$$

(..........., ...)

(2 marks)

(b) (2, 6) and (8, 10)

$$\text{midpoint} = \left(\frac{2+8}{2}, \frac{6+10}{2}\right)$$

$$= \left(\frac{\ldots\ldots}{2}, \frac{\ldots\ldots}{2}\right)$$

(..........., ...)   **(2 marks)**

(c) (4, 9) and (−4, 11)

$$\text{midpoint} = \left(\frac{\ldots\ldots + \ldots\ldots}{2}, \frac{\ldots\ldots + \ldots\ldots}{2}\right)$$

$$= \left(\frac{\ldots\ldots}{2}, \frac{\ldots\ldots}{2}\right)$$

(..........., ...)   **(2 marks)**

(d) (5, −5) and (−11, 1)

(..........., ...)   **(2 marks)**

3 Look at the diagram. *P* is the midpoint of both lines *OS* and *QR*, and *R* is the coordinate (10, 5).

Guided

(a) Work out the coordinates of *Q*.

    ....   **(3 marks)**

(b) Work out the coordinates of *S*.

    ....   **(3 marks)**

Gradients of lines

1 The gradient *m* (measure of the slope, not the length) of a line is defined as its vertical height (*y*) divided by its horizontal length (*x*).

  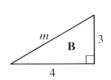

slope $(m) = \dfrac{\text{vertical distance}}{\text{horizontal distance}} = \dfrac{\text{distance in } y}{\text{distance in } x}$

$(m) = \dfrac{y}{x}$

(a) Work out the value of gradient *m* in triangle **A**.

$m = \dfrac{4}{\dots\dots} = \dots\dots\dots$ **(2 marks)**

(b) Work out the value of gradient *m* in triangle **B**.

$m = \dfrac{\dots\dots}{\dots\dots} = \dots\dots\dots$ **(2 marks)**

2 Look at the lines **A** and **B**. Draw a right-angled triangle under each one, to work out the gradient *m* of each.

(a) Gradient *m* of **A**:

$m = \dfrac{\dots\dots}{\dots\dots} = \dots\dots\dots$ **(2 marks)**

(b) Gradient *m* of **B**:

$m = \dfrac{\dots\dots}{\dots\dots} = \dots\dots\dots$ **(2 marks)**

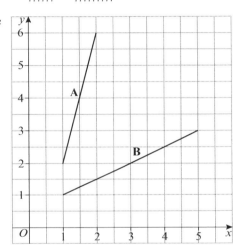

1 Let the line be the hypotenuse of a right-angled triangle.
2 Complete the right-angled triangle.
3 Find the distance in *y* and the distance in *x*.
4 Use the formula for calculating a gradient.
5 Remember to use negative signs if the distance is a negative direction.

3 Work out the value of the gradient *m* on the line shown. (The point (1, 1) is where the right angle can be placed to form a right-angled triangle.)

$m = \dfrac{\dots\dots}{\dots\dots} = \dots\dots\dots$

(2 marks)

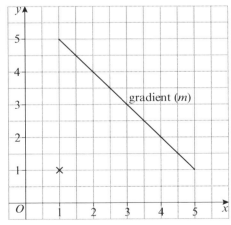

Straight-line graphs 1

1 (a) Complete the table for the missing x and y values.

$y = 2x - 1$

x	−1		2
y		−1	

(2 marks)

(b) Plot the three (x, y) coordinate points from your table and draw a straight line through them. Do not start the line at the first point and end at the last point. Draw the line through the points so that both ends touch the edge of the grid.

(2 marks)

(c) Is it true that the coordinate $(−2, −6)$ will also lie on the line? **(1 mark)**

(d) If the grid were larger, do you think the coordinates $(12, 23)$ would lie on the line? Explain your reasoning.

> Check that the coordinates satisfy the equation $y = 2x - 1$.

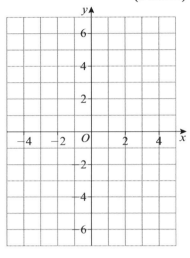

....

.... **(2 marks)**

2 Draw the graph of $x - y = 4$. Choose three suitable values of x to form three pairs of coordinates.

> Guided

$x - y = 4$

x			
y			

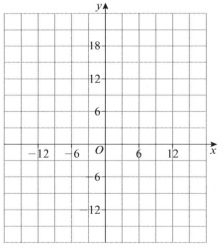

(2 marks)

3 Work out the equation of this straight line.

x			
y			

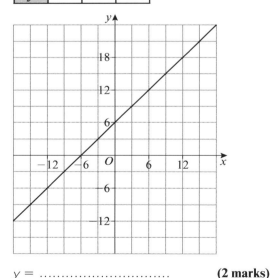

$y = $ .... **(2 marks)**

> Draw a triangle using the line as the hypotenuse – choose easy points to plot and measure from. Use the vertical and horizontal sides of the triangle to find the gradient.

Straight-line graphs 2

4 Work out the equation of a straight line with

(a) gradient of 2 and passing through the point (2, 3)

> $y = 2x + c$
> $(3) = 2(2) + c$
> $3 = 4 + c$
> $y = ...$

> Substitute the gradient (m) and the values of x and y given so that we have an equation in terms of c.

(2 marks)

(b) gradient of −1 and passing through the point (2, 3)

> $y = ...$

(2 marks)

5 Find the equation of the straight line that passes through the points (1, 3) and (3, 5).

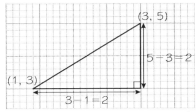

> $m = \dfrac{\text{difference in } y \text{ values}}{\text{difference in } x \text{ values}}$
>
> $m = \dfrac{2}{2} = 2 \div 2 = 1$
>
> $y = mx + c$
>
> $3 = 1 \times 1 + c$
>
> $3 = 1 + c$
>
> $c = 2$
>
> $y = ...$

> You can use this value of $m = 1$ and substitute in one of the pairs of values for x and y.

(3 marks)

6 Find the equation of line **A**, given that both lines are parallel.

PROBLEM SOLVED!

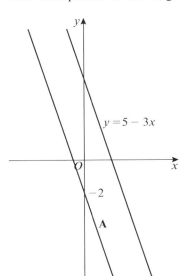

> You will need to use problem-solving skills throughout your exam – **be prepared!**

> $y = 5 - 3x$

> $y = ...$

(2 marks)

Real-life graphs

1   This graph can be used to convert between miles and kilometres.

   (a)   Use the graph to convert 30 miles to kilometres.

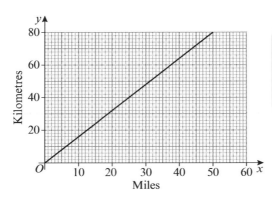

   > Draw a vertical line from 30 on the miles scale until you touch the line of the graph; then, from that point, draw a horizontal line until you touch the km scale.

   30 miles = ...   **(1 mark)**

   (b)   Convert 60 kilometres to miles

   60 km = ...   **(1 mark)**

   (c)   The distance between London and Calais is 188 kilometres. The distance between London and Bournemouth is 108 miles. Which town is closer to London?

   > Divide both distances by 4 so that you can use the graph.

   ....   **(3 marks)**

2   A construction company sells houses on a plot of land based on the number of square metres each house takes up. The graph is used to work out both cost and area.

   (a)   A plot of land with an area of 48 m² was sold. How much did it cost?

   > You will need to use problem-solving skills throughout your exam – **be prepared!**

   £ ....   **(1 mark)**

   (b)   A plot of land costs £90 000. What is the area of the plot?

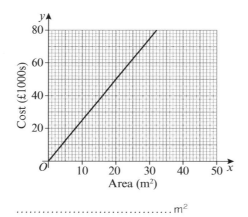

   > The graph can be used with the £30 000 point on the scale, as this number is a factor of £90 000. You will need to multiply your answer by 3 because 30 000 is three times smaller than 90 000.

   .... m²   **(3 marks)**

Distance–time graphs

1 Rachael went to her local café on her bicycle, then cycled back home.

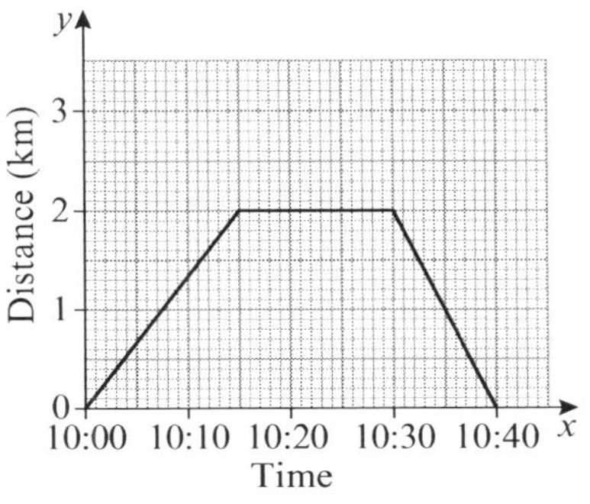

(a) What time did she start her journey?

.... **(1 mark)**

> This is the left-most point of the graph on the time axis.

(b) What is the distance of the complete cycle ride?

.... **(1 mark)**

> Add up the total vertical distance travelled on the graph.

(c) How long was Rachael in the café?

.... **(1 mark)**

> The flat section of graph shows the time when Rachael was stationary.

(d) What was her average speed from home to the café?

> (s) is found by dividing distance (d) by time (t) taken. You can use the formula $s = \frac{d}{t}$.

.... km/h **(2 marks)**

2 Gurjit's 50 km drive to see her uncle took 45 minutes. She spent an hour and a half with him before returning home. Her journey back took 1 hour.

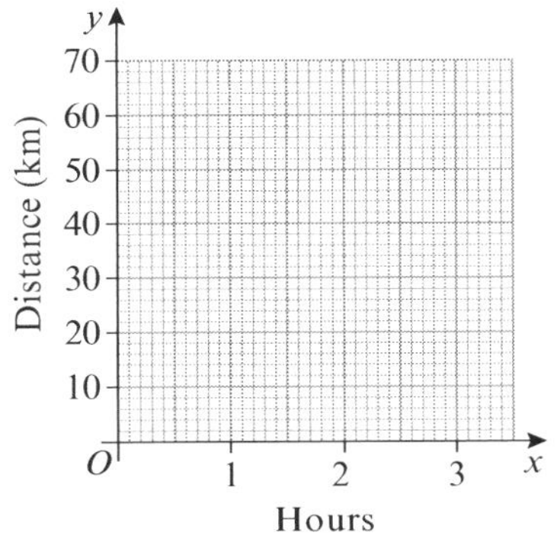

(a) Complete the distance–time graph.

> The graph will start at (0, 0) and then have a straight line to the point with coordinates 50 km and 45 minutes.

(2 marks)

(b) If Gurjit returned home at 18:20, what time did she start the whole journey?

> Use the graph to work out the total time and then count backwards from 18:20. Remember there are 60 minutes in an hour, not 100.

.... **(2 marks)**

(c) What time did Gurjit arrive at her uncle's house?

.... **(1 mark)**

Rates of change

1 Here are silhouettes of 4 vases. The graphs show the rate of change of the depth of water as each vase is filled (d = depth, t = time). Draw a line to link each vase with its correct graph.

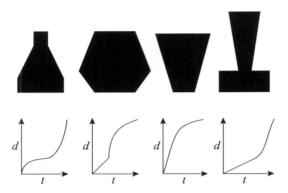

> A straight line section on the graph indicates a constant surface area is being filled. A curved section indicates the surface area of the water is either getting smaller or larger over time.

(4 marks)

2 The graph shows the rate at which a 45 000 litre swimming pool was filled.

(a) Write down how much water was in the pool at the start.

.... litres   **(1 mark)**

(b) Work out the gradient of the line.

....   **(2 marks)**

(c) Explain what the gradient of the line means in real terms.

....
....
....   **(2 marks)**

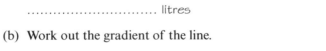

3 Here is a velocity–time graph of a motorbike.

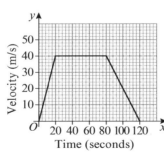

> The rate of change in velocity means acceleration or deceleration.
> Rate of change = gradient = $\dfrac{\text{vertical distance}}{\text{horizontal distance}}$

(a) Calculate the rate of change in the first 20 seconds.

.... m/s²   **(2 marks)**

(b) Explain what was happening between 20 and 80 seconds.

....   **(1 mark)**

(c) What is the change in velocity in the last 40 seconds?

= .... − ....

> Subtract the final velocity from the velocity at the start of the 40-second period.

.... m/s²   **(2 marks)**

Expanding double brackets

1 Expand the brackets and simplify

(a) $(x + 1)(x + 2)$

$= x(x + 2) + 1(x + 2)$

$= x^2 + 2.......... + x + ...$

$x^2 + ... + ...$

> This method helps to reduce mistakes as the terms in one bracket multiply the other bracket in turn.

(2 marks)

(b) $(x + 2)(x + 4)$

$= x(.......... + ...)$

$+1(.......... + ...)$

$= x^2 + 2.......... + 4.......... + ...$

$= x^2 + ... + ...$ **(2 marks)**

(c) $(x + 3)(x + 4)$

$= .... (.......... + ...)$

$+ .... (.......... + ...)$

$= x^2 + ... + ... + ...$

$= .... + ... + ...$ **(2 marks)**

(d) $(x + 3)(x - 1)$

.... **(2 marks)**

(e) $(x - 3)(x - 2)$

.... **(2 marks)**

2 Expand and simplify

(a) $(2x + 1)(3x + 2)$

$= 2x(3x + 2) + 1(3x + 2)$

$= 6x^2 + 4.......... + ... + ...$

$= 6x^2 + ... + ...$

> You could also have written $3x(2x + 1) + 2(2x + 1)$. This would give exactly the same result.

 You will need to use problem-solving skills throughout your exam – **be prepared!**

(2 marks)

(b) $(5x - 3)(x + 2)$

$= x(5x - 3) + ...(...... - ......)$

$= 5x^2 - 3.......... + ... - ....$

$= 5x^2 + ... - ....$ **(2 marks)**

(c) $(3x - 2)(3x - 1)$

$= 3 \times 3..........^2 - .... - ....$

$+ ....$

$= ....x^2 - .... + ...$ **(2 marks)**

(d) $(x + 5)^2$

$= (x + 5)(x + 5)$

.... **(2 marks)**

(e) $(2x - 7)^2$

.... **(2 marks)**

3 If n is a whole number, show that $(n + 1)^2 - n^2$ is always an odd number.

> Expand the brackets and simplify. What remains?

....

.... **(2 marks)**

4 If a and b are integers, is it true that $a^2 + b^2 - (a - b)^2$ produces both odd and even numbers? You must show your working.

.... **(4 marks)**

5 A triangle has the dimensions shown. Explain why it is either true or false that the area of the triangle $= \dfrac{3x^2 + 7x + 2}{2}$.

.... **(4 marks)**

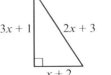

$3x + 1$ $2x + 3$

$x + 2$

Quadratic graphs

1   Here is a table of coordinates for $y = x^2 - 1$.

x	–3	–2	–1	0	1	2	3
y	8			–1			

(a) Complete the table.

> Substitute each value of x in the table into the equation to find its corresponding y-value.
> For example, when $x = -2$
> $y = (-2) \times (-2) - 1 = 4 - 1$
> So, $y = 3$.

(2 marks)

(b) Use the information in the table to draw the graph of $y = x^2 - 1$.

(2 marks)

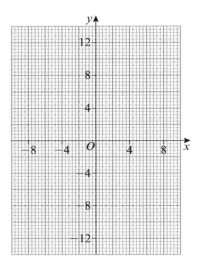

(c) Write down the coordinates at the turning point of the curve.

> The turning point is where the curve changes to the opposite direction.

....   **(2 marks)**

(d) Use your graph to find an approximate value of y when $x = 1.5$. Check the accuracy of your answer by substituting this value of x into the equation.

....   **(2 marks)**

2   Here is a table of coordinates for $y = x^2 - 2x - 3$.

x	–3	–2	–1	0	1	2	3
y	12			–3			

(a) Complete the table.   **(2 marks)**

(b) Use the information in the table to draw the graph of $y = x^2 - 2x - 3$.   **(2 marks)**

(c) Write down the coordinates at the turning point of the curve.

....   **(2 marks)**

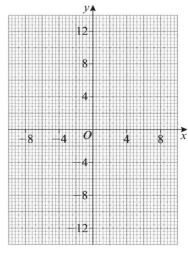

Using quadratic graphs

1 Here is a table of coordinates for $y = x^2 + x + 1$.

x	–3	–2	–1	0	1	2	3
y			1			7	

(a) Complete the table. **(2 marks)**

(b) Using the information in the table, draw the
graph of $y = x^2 + x + 1$. **(2 marks)**

(c) Use the graph to estimate the minimum value of y.

> The graph can only be used for an estimation
> as it is unusual for anyone to draw a completely
> accurate quadratic graph.

$y = $ .... **(1 mark)**

(d) Work out the exact value for y at the minimum value.

> The minimum value of y is found at the turning
> point. The exception is where the curve is the
> other way up, in which case the turning point shows
> the maximum value of y. This occurs in quadratic
> equations that contain a negative x^2 term.

$y = $ .... **(2 marks)**

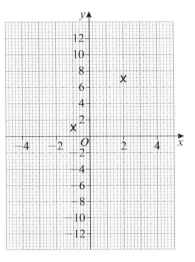

2 Here is a table of coordinates for $y = 4 - 2x - x^2$.

x	–4	–3	–2	–1	0	1	2
y							

(a) Complete the table. **(2 marks)**

(b) Draw the graph of this equation on the grid.

  (2 marks)

(c) Give an estimate for the coordinates of the turning point.

.... **(1 mark)**

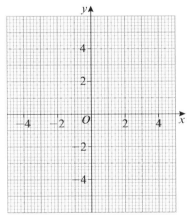

Factorising quadratics

 1 Find the value of x in each quadratic equation.

(a) $x^2 = 16$

> You need to find the square roots of 16.

$x = $ .... and .... **(1 mark)**

(b) $x^2 = \frac{1}{4}$

$\frac{1}{4} = \frac{1}{2} \times \frac{1}{2} = \left(-\frac{1}{2}\right) \times \left(-\frac{1}{2}\right)$

$x = $ .... and ....

(2 marks)

(c) $2x^2 = 72$

> Dividing both sides by 2 gives
> $x^2 = 36$
> $36 = 6 \times 6 = (-6) \times (-6)$

$x = $ .... and .... **(1 mark)**

 2 Factorise

> Find factors of both terms on the right-hand side and place outside brackets.

(a) $y = x^2 + 6x$

$y = x(x + $ ....$)$ **(1 mark)**

(b) $y = x^2 + 12x$

$y = x($ .... $+ 12)$ **(1 mark)**

(c) $y = x^2 + 2x$

$y = x($ .... $+$ ....$)$ **(1 mark)**

(d) $y = x^2 - x$

$y = $ .... **(1 mark)**

(e) $y = ax^2 + bx$

$y = $ .... **(1 mark)**

3 Factorise

(a) $y = x^2 + 4x + 3$

> Find a factor pair of the constant which, when added, sum to the coefficient or multiplier of x.

$y = x^2 + x + 3x + 3$
$y = (x^2 + x) + (3x + 3)$
$y = x(x +1) + 3(x + 1)$

> Now put a bracket around the first two terms and another around the last two terms.

$y = (x + 1)(x + $ ....$)$

> If the contents of both brackets are the same, then that is a factor of the original expression. In this case $(x + 1)$ is a factor. The other factor is the sum of the terms multiplying the brackets.

(2 marks)

(b) $y = x^2 + x - 12$

$y = (x - 3)($ .... $+$ ....$)$

> Where the constant is a negative, you will need a factor pair that you can use for a subtraction to get the right x coefficient.

(2 marks)

(c) $y = x^2 - 5x + 6$

$y = ($ .... $-$ ....$)($ .... $-$ ....$)$ **(2 marks)**

 4 Factorise

(a) $y = x^2 - 144$

....

> If the expression is just x^2 minus a constant that is a square number, this is called 'the difference of 2 squares'.

(2 marks)

(b) $y = x^2 - 1$

....

> In general, these expressions are in the following form, where a is a square number:
> $y = x^2 - a$

(2 marks)

(c) $y = 2x^2 - 18$

....

> You can always factorise these expressions in this way: $y = (x + \sqrt{a})(x - \sqrt{a})$.

(2 marks)

Quadratic equations

1 Solve

  (a) $x^2 - 3x = 0$

> Factorise first.

You will need to use problem-solving skills throughout your exam – **be prepared!**

    $x(x - 3) = 0$

    $x = 0$ or $x =$ ....

> When you multiply two numbers and the result is 0, at least one of the numbers must be equal to 0.

(2 marks)

  (b) $x^2 + 3x = 0$

    $x(x + 3) = 0$

    $x =$ .... or $x =$ ....

(2 marks)

  (c) $x^2 - 5x = 0$

    $x(x -$ ....$) = 0$

    $x =$ .... or $x =$ ....

(2 marks)

  (d) $2x^2 + 8x = 0$

> Factorise number parts first.

    ....

(3 marks)

2 Solve

  (a) $x^2 + x - 2 = 0$

You will need to use problem-solving skills throughout your exam – **be prepared!**

> Factorise first.

    $x^2 + x - 2 = (x - 1)(x + 2)$

    $(x - 1)(x + 2) = 0$

> Re-write the equation.

    $x =$ .... or $x =$ ....

> The contents of at least one set of brackets must be equal to zero. So, either $x - 1 = 0$ or $x + 2 = 0$.

(2 marks)

  (b) $x^2 - 5x + 4 = 0$

    $(x - 4)(x - 1) = 0$

    $x =$ .... or $x =$ ....

(2 marks)

  (c) $x^2 + 4x - 21 = 0$ (d) $x^2 + 9x + 18 = 0$

    $(x -$ ....$)(x +$ ....$) = 0$

    $x =$ .... or $x =$ .... **(2 marks)**   .... **(2 marks)**

3 Use the 'difference of two squares' to solve

  (a) $x^2 - 16 = 0$ (b) $x^2 - 100 = 0$

    .... **(2 marks)**   .... **(2 marks)**

  (c) $x^2 - 25 = 0$ (d) $3x^2 - 12 = 0$

    .... **(2 marks)**   .... **(2 marks)**

4 The product of two consecutive positive even numbers is 440. Construct a quadratic equation to find the two numbers in question.

> You could call the consecutive even numbers $x + 1$ and $x - 1$. What expression are they the factors of? What is the whole expression equal to?

> 'Product' means the multiple of the two numbers.

    ....

(3 marks)

Cubic and reciprocal graphs

1 Here is a table of coordinates for $y = x^3 - 3x + 1$.

x	-2	-1	0	1	2
y		3			3

(a) Complete the table.   **(2 marks)**

(b) Complete the graph of $y = x^3 - 3x + 1$.   **(2 marks)**

(c) Write down the coordinates of the turning points.

> A turning point is where the graph changes to the opposite direction.
> Use the graph to estimate the turning point, then put the x- and y-coordinates at this point into the equation to see whether your estimate is accurate.

(............, ...)(............, ...)   **(2 marks)**

(d) Draw the line $y = -3$ on the same graph paper, then use the graph to estimate the value of x when $x^3 - 3x + 1 = -3$.

> The line $y = -3$ passes through the point $(0, -3)$ on the y-axis and is parallel to the x-axis. You need to find the place where this line cuts the cubic graph.

....

(2 marks)

2 Look at the graphs. Match each one with a possible equation.

Gu~~i~~ed

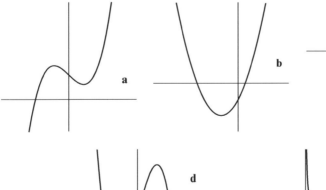

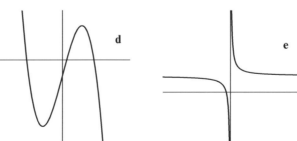

(i) $y = \frac{1}{x} + 2$

.... **(1 mark)**

(ii) $y = x^3 - x + 1$

.... **(1 mark)**

(iii) $y = 2x + 1 - x^2$

.... **(1 mark)**

(iv) $y = 4x - x^3 - 1$

.... **(1 mark)**

(v) $y = x^2 + 2x - 1$

.... **(1 mark)**

Simultaneous equations

1 Solve the simultaneous equations.

(a) $x + 2y = 4$ [1]

 $x - y = 1$ [2]

 $(x - x = 0) + (2y - (-y)) = 4 - 1$

 $3y = 3$ [3]

 $y = \dots\dots\dots$ **(2 marks)**

 $x + 2(\dots\dots) = 4$

 $x = \dots\dots\dots$ **(1 mark)**

 $(\dots\dots) - (\dots\dots) = \dots\dots$

> Label each equation with a number or letter.

> To eliminate one variable (x), subtract the terms in [2] from the terms in [1] and label the result [3].

> Now to find x, substitute the value of y into equation [1].

(b) $3x + 2y = 11$ [1]

 $2x - 5y = 20$ [2]

 $\dots\dots\dots\dots y = - \dots\dots\dots$

 $y = \dots\dots\dots$

 $3x + 2 (\dots\dots) = 11$

 $\dots\dots\dots\dots\dots\dots\dots\dots\dots\dots$ **(3 marks)**

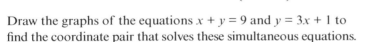

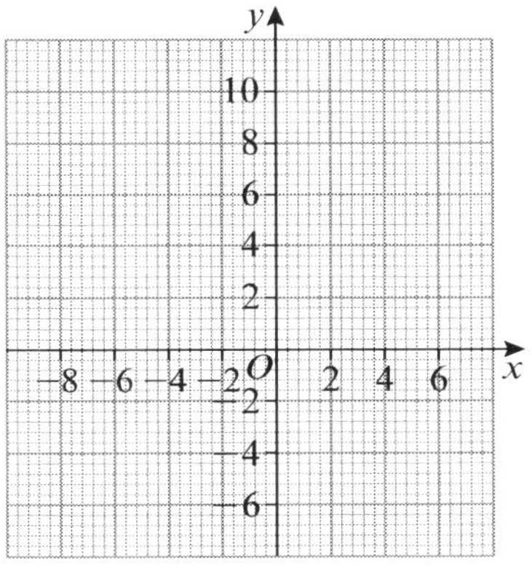

2 Draw the graphs of the equations $x + y = 9$ and $y = 3x + 1$ to find the coordinate pair that solves these simultaneous equations.

> For each line, choose three values of x and find their matching y values, then plot and draw a line all the way through the points.

$x + y = 9$

x	0	2	4
y	9	7	5

(0, 9) (2, 7) (4, 5)

> Now create a table for $y = 3x + 1$.

> Plot both graphs on the grid. The point where they cross will be the solution to the simultaneous equations $x + y = 9$ and $y = 3x + 1$.

$x = \dots\dots\dots$ $y = \dots\dots\dots$ **(4 marks)**

3 Draw the graphs of the equations $2x + y = 7$ and $x + y = 6$ to find the coordinate pair that solves these simultaneous equations.

$x = \dots\dots\dots$ $y = \dots\dots\dots$ **(4 marks)**

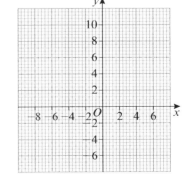

Rearranging formulae

1 In the formula $3ab + 5 = c$, the subject is c.

(a) Rearrange the formula to make a the subject.

$3ab + 5 = c$

$3ab + 5 - 5 = c - 5$

$3ab = c - 5$

$a = \dots\dots\dots\dots\dots$

> Start by subtracting 5 from both sides.

> Think about what you must divide by in order to leave a by itself. Do this to both sides of the equals sign.

(2 marks)

(b) If $b = 4$ and $c = 41$, use the formula to find the value of a.

$a = \dfrac{\dots\dots}{\dots\dots}$

> Write the formula where a is the subject.

(2 marks)

$a = \dots\dots\dots\dots\dots$

(2 marks)

> Substitute the values of b and c into the formula, then calculate the value of a.

2 Make x the subject of the formula $3x - 2y = 10$.

> Add a term to both sides to leave only $3x$ on the left, then divide both sides by the same number to make x the subject.

$x = \dots\dots\dots\dots\dots$

(2 marks)

3 Make x the subject of the formula $-2x + 8y = 14$.

⟩ Guided ⟩

$\dots\dots\dots\dots\dots$

(2 marks)

4 Sheila rears hens on an organic farm. She uses the formula below to work out the price, in pence (P), of the eggs she sells.

$P = OH + LW$

(a) Rearrange the formula for O, which tells Sheila the cost impact of using organic chicken-feed.

$P - LW = \dots\dots\dots\dots\dots$

$O = \dfrac{\dots\dots\dots\dots\dots}{\dots\dots\dots\dots}$

> Divide both sides by a letter to leave O on its own.

(2 marks)

(b) Calculate O if $H = 40$, $W = 20$, $L = 2$ and $P = 120$.

$O = \dfrac{\dots\dots - (\dots\dots \times \dots\dots)}{\dots\dots\dots\dots} =$

> Substitute these values into your rearranged formula.

$O = \dots\dots\dots\dots\dots\dots$

(2 marks)

Using algebra

1 Rod sells apples on the market. He has 12 boxes to sell and each box contains n apples. He checks his boxes for bad apples and throws 15 away. He now has 345 apples to sell.

(a) Write an equation that explains this information.

| Each box has n apples and there are 12 boxes. |

Rod has 12 .... apples at the start.

| He throws away 15 of the apples. |

After throwing away the bad apples Rod has

| He is left with 345 apples to sell. |

12 × .... − .... apples.

Rod can say 12 × .... − .... = ... apples.

This can be written as the equation ....   **(2 marks)**

(b) Work out how many apples were in each box.

| Solve the equation for n. |

$n =$ ....   **(2 marks)**

2 The diagram shows a plot of land measured in metres. Work out the value of n if the total perimeter of the plot is 850 m.

| Form an equation. First sum the lengths of the sides, then collect like terms. Set the total equal to 850. Then solve the equation. |

| You will need to use problem-solving skills throughout your exam – **be prepared!** | |

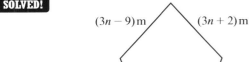

$(3n − 9)$ m $(3n + 2)$ m

$(2n + 5)$ m $(2n − 1)$ m

$(2n + 13)$ m

$(3n − 9) +$ .... $+$ .... $+$ .... $+$ .... $= 850$

$n =$ ....   **(3 marks)**

3 The rectangle and triangle shown have the same perimeters. Work out the length of side y.

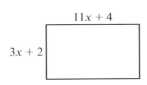

$11x + 4$

$3x + 2$

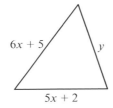

$6x + 5$ y

$5x + 2$

$y =$ ....   **(3 marks)**

4 The area of the triangle is exactly half that of the rectangle. Work out the base length (b) of the triangle.

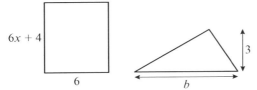

$6x + 4$

6

3

b

$b =$ ....   **(3 marks)**

Identities and proof

1 Show that $(3n + 2)^2 + (3n - 2)^2 \equiv 2(9n^2 + 4)$

> Multiply out the brackets first.

$(3n + 2)^2 = (3n + 2)(3n + 2) = 9n^2 + \ldots\ldots\ldots + \ldots\ldots\ldots + \ldots\ldots\ldots$

$(3n - 2n)^2 = (3n - 2n)(3n - 2n) \ldots\ldots\ldots - \ldots\ldots\ldots - \ldots\ldots\ldots + \ldots\ldots\ldots$

> Now, collect like terms and simplify the result, remembering to fully factorise.

(3 marks)

2 Prove that the sum of two consecutive integers is always an **odd** number.

> Let the first number be n, therefore, the second number is $n + 1$.
> Add them together then simplify.

$n + (n + 1) = \ldots\ldots\ldots$

> Decide whether each term is odd or even and from there make your conclusion about the statement.

....

.... **(2 marks)**

3 In the identity $2(x + 1)(x + 2) \equiv Ax^2 + Bx + C$, prove that for any value of x, $A + B + C = 12$

> Guided

(3 marks)

4 If $x = 20$, use the identity $(x + y)(x - y) \equiv x^2 - y^2$ to solve 22×18

> Guided

(3 marks)

5 Show that the sum of the first 3 numbers from 5 consecutive numbers and the sum of the last 3 consecutive numbers from the same 5 are both divisible by 3.

> PROBLEM SOLVED!

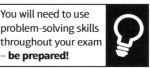

You will need to use problem-solving skills throughout your exam – **be prepared!**

> Let the numbers in the sequence be labelled n, $(n + 1)$, $(n + 2)$, $(n + 3)$, and $(n + 4)$. Use these general terms in a fraction where the numerator is the sum of the first three terms and the denominator is the sum of the last three terms. Simplify and factorise numerator and denominator.

$$\frac{n + (n + 1) + \ldots\ldots\ldots}{\ldots\ldots\ldots + \ldots\ldots\ldots + (n + 4)}$$

(4 marks)

Problem-solving practice 1

1 Amy and Sid wanted to transform their garden and hired a digger to do the job. The cost of hiring the digger was £42.75 for the first day and £23.60 for each day after that. They were charged £113.55 for the total hire. How many days did Amy and Sid hire the digger for?

.... **(3 marks)**

2 Two coordinates of a right-angled triangle are given as $(-6, 2)$ and $(6, 2)$. The third coordinate contains no negative values and the area of the triangle is 48 square units.

(a) Work out the missing coordinate.

.... **(2 marks)**

(b) Write down what the missing coordinate would be if both x and y values were negative.

.... **(2 marks)**

3 An engineering firm costs each production job using the formula $C = 65h + 235$, where C = total cost, h = number of hours the job takes and 235 is the set-up cost.

(a) Calculate the cost of a production job that takes 8 hours to complete.

.... **(2 marks)**

(b) One job worked out at a cost of £397.50. Calculate how long this job took to complete.

.... **(2 marks)**

4 A group of six people go to the theatre and sit next to each other in a row where the seats are numbered consecutively. The sum of their seat numbers is 555. What are the individual seat numbers of the group?

.... **(3 marks)**

5 Tickets are priced for adults and children at the cinema. Tickets for a family of 3 adults and 2 children cost £46 and tickets for a family of 2 adults and 3 children cost £44.

Find the cost of

(a) an adult ticket.

.... **(2 marks)**

(b) a child ticket.

.... **(2 marks)**

Had a go ☐ **Nearly there** ☐ **Nailed it!** ☐

Problem-solving practice 2

6 Work out the value of n.

$$2^{n-5} = \frac{2^{15} \times 2^{-3}}{2^6}$$

(2 marks)

7 The circumference of the Earth at the equator is approximately 4.003×10^4 km. The height of the Lincoln Cathedral spire is approximately 160 m. How many Lincoln Cathedral spires could be laid end to end around the equator? Write your answer to 3 significant figures.

(2 marks)

8 Here are the first 5 terms in a linear sequence.

$$3 \quad 5 \quad 7 \quad 9 \quad 11$$

(a) Work out the nth term of this sequence.

.... **(2 marks)**

(b) Work out the number of the term in the sequence which has the value 71.

.... **(1 mark)**

9 The diagram shows a triangle with vertical height $3x + 8$ and a base length of $4x - 2$. Show that the formula for the area (A) of this triangle is $A = 6x^2 + 13x - 8$.

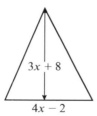

$3x + 8$

$4x - 2$

(3 marks)

10 The rectangle shown has a perimeter of p cm and an area of $p + 23$ cm^2.

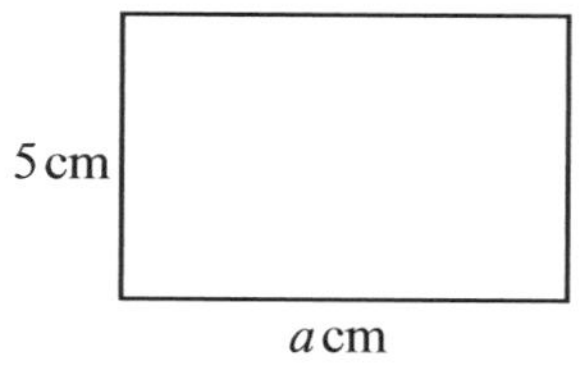

(a) Work out the perimeter of the rectangle.

....

5 cm

(3 marks)

(b) Work out the area of the rectangle.

....

a cm

(3 marks)

54

Percentages

1 Find

    (a) 8% of 20

> Percentages, decimals and fractions are interchangeable. Use the most efficient in a problem.

$$= \frac{8}{100} \times 20 = 8 \times \frac{20}{100} = \frac{160}{100}$$

> Convert 8% to a fraction then multiply.

    = ... **(3 marks)**

    (b) 7% of 230

        = ... **(2 marks)**

2 Express as a percentage

    (a) 27 out of 72

> Write as a fraction.

$$\frac{27}{72}$$

> Simplify the fraction.

$$\frac{27}{72} = \frac{3}{8}$$

> Convert the simplified fraction to a percentage (this can be done by converting to a decimal first).

    = ... **(1 mark)**

    (b) 34 out of 40

        = ... **(1 mark)**

3 A lawnmower costs £76 plus 20% VAT.

    (a) What is the VAT cost?

> Find 20% of £76.

    .... **(2 marks)**

    (b) What is the total cost of the lawnmower?

> Add the result of part (a) to £76.

    .... **(2 marks)**

4 The table shows the different types of questions on a mathematics exam.

Topic	Number of questions
Number	26
Algebra	20
Geometry	19
Statistics	10
Probability	5

    (a) What percentage of the questions were on Probability?

    ....

    (b) What percentage of the questions were on Statistics and Algebra?

    ....

    (c) Jaden says that the total of Number and Algebra questions amounted to 46% of the paper. Mikita says it is more like 58%. Show which person is correct.

    ....

Fractions, decimals and percentages

1 Write the percentages as fractions in their simplest forms.

(a) 42%

$$42\% = \frac{42}{100}$$

$$\frac{42}{100} = \frac{\ldots\ldots\ldots}{50}$$

> Convert 42% to a fraction. Remember per cent literally means 'per one hundred'.

> Simplify by dividing both parts by 2.

.... **(2 marks)**

(b) 56%

$$56\% = \frac{\ldots\ldots\ldots}{100} = \frac{\ldots\ldots\ldots}{\ldots\ldots}$$

.... **(2 marks)**

(c) 96%

$$96\% = \frac{\ldots\ldots\ldots}{\ldots\ldots} = \frac{\ldots\ldots\ldots}{\ldots\ldots}$$

(2 marks)

2 Order the following numbers starting with the smallest.

(a) $\frac{2}{3}$, 0.6, 67%

$$\frac{2}{3} = 0.667 \ (3 \text{ s.f.})$$
$$67\% = 0.670 \ (3 \text{ s.f.})$$

The order is ...., ..., ...

(2 marks)

(b) 38%, $\frac{29}{75}$, 3.8

$$\frac{29}{75} = \ldots\ldots\ldots$$
$$38\% = \ldots\ldots\ldots$$

The order is ...., ..., ...

(2 marks)

(c) $\frac{3}{20}$, 0.151, $\frac{4}{21}$

.... **(2 marks)**

3 There are 160 Year 11 pupils in a school. 30% study only French, 35% study only German and $\frac{3}{8}$ study both languages.

(a) Work out how many students study both French and German.

.... **(2 marks)**

(b) Calculate the difference between the number of students who study only French and those who do not study French.

.... **(2 marks)**

4 Jack and Nisha both keep chickens. In one month, Jack's chickens ate 70% of a 3 kg bag of corn. Nisha's birds ate $\frac{3}{5}$ of a 4 kg bag. Calculate whose chickens ate more corn that month. You must show your working.

.... **(3 marks)**

Percentage change 1

1 Work out the following percentage increases.

(a) Increase 72 by 15%.

115% of 72 = 72 × 1.15

....

> If 72 represents 100% then by increasing it by 15% we are finding 115% of 72.

(2 marks)

(b) Increase 40 by 12%.

112% of 40 = ... × 1.12

.... **(2 marks)**

(c) Increase 27 by 8%.

108% of 27 = ... × ....

.... **(2 marks)**

2 Work out the following percentage increases.

(a) From 32 to 40

$\frac{32}{32} = 1 = 100\%$ therefore, $\frac{40}{32} = \frac{5}{4} = 1.25$

$1.25 - 1 = 0.25$

> Subtract 100% and multiply by 100 to get the percentage increase.

Percentage increase = 0.25 × 100 = ...% **(2 marks)**

(b) From 32 to 90. Give your answer to 2 decimal places.

Percentage increase = ...% **(2 marks)**

3 Work out the following percentage decreases (to 2 decimal places).

(a) From 56 to 30

$\frac{56}{56} = 1 = 100\%$ therefore $\frac{30}{56} = \frac{15}{28} = 0.$ .... = ...%

Percentage decrease = ...% **(2 marks)**

(b) From 56 to 8

Percentage decrease = ...% **(2 marks)**

4 Esther and Kenji both received a 3% pay rise. Before their increases, Esther was earning £32 600 and Kenji was earning £30 970. (Round to 1 decimal place where necessary.)

> You will need to use problem-solving skills throughout your exam – **be prepared!**

(a) Calculate Kenji's new salary as a percentage of Esther's new salary.

.... **(4 marks)**

(b) State whether the percentage difference in their salaries is the same as it was before their pay increases, or whether it has changed. Give a reason for your answer.

....

.... **(2 marks)**

Percentage change 2

5 Two online concert ticket providers are selling tickets for a rock concert.

If tickets are advertised at £55 each, which ticket provider is cheaper?

Ticket Shop 1

$55 \times \dfrac{\text{..............}}{100} = $ ....

£55 + £ .... + £ .... = £ ....

Ticket Shop 2

$55 \times \dfrac{\text{..............}}{100} = $ ....

£55 + £ .... + £ .... = £ ....

.... is cheaper.   **(4 marks)**

6 Denis needs to buy floor tiles for his bathroom. He has worked out he needs 40 of them. There are two shops where he can buy the tiles.

Denis thinks A&P Stores offers the better deal. Work out whether he is right.

> First, find the total cost of 40 tiles at A&P Stores, then find 12% of this figure and subtract it from the total.
>
> Next, find the total cost of 40 tiles at Fixnfit, then add 20% of this cost to get the total price to pay.

....

....   **(4 marks)**

7 Mikey needs to buy a new tumble dryer. The model he is looking for is sold by two online stores.

Which store offers the better deal on this tumble dryer? You must show your working.

....   **(4 marks)**

Had a go ☐   Nearly there ☐   Nailed it! ☐

Ratio 1

1 Use the highest common factor (HCF) of each number in the following ratios to write them in their simplest form.

(a) 15 : 75

15 = 3 × 5

75 = 3 × 5 × 5

HCF (15, 75) = 3 × 5 = 15

Ratio is equal to $\frac{15}{15} : \frac{75}{15}$

So 15 : 75 = ... : ....

(2 marks)

(b) 24 : 32

24 = 2 × 2 × 2 × 3

32 = 2 × 2 × 2 × 2 × 2

HCF (24, 32) = 2 × 2 × 2 = ...

$24 : 32 = \frac{............}{8} : \frac{............}{............}$

24 : 32 .... : ....

(2 marks)

(c) 48 : 100

....

(2 marks)

2 (a) Divide £72 in the ratio 5 : 4

> The sum of the numbers used in the ratio is the total number of parts.

Total parts = ... + ...

1 part is equal to £72 ÷ ....

5 parts = ... × .... = £............

4 parts = ... × .... = £............

The two amounts are £............ and £............

> You will need to use problem-solving skills throughout your exam – **be prepared!**

(2 marks)

(b) Divide 42 kg in the ratio 5 : 9

5 + 9 = ...

42 ÷ 14 = ...

.... × .... = ...

.... × .... = ...

The two amounts are .... kg and .... kg

> To find one part, sum the ratio parts then divide the number to be split by the total number of parts. Multiply this by how many parts you need.

(2 marks)

(c) Divide £121 in the ratio 8 : 3

....

(2 marks)

3 Two groups of people travelled in two cars from Bristol to Penzance, which is 190 miles. In car A the ratio of miles driven by the people travelling was Millie : Niall : Ortensia = 5 : 6 : 8. In car B, the ratio was Robbie : Saira : Tannu = 2 : 3 : 5.

(a) Nikki says that Tannu and Millie drove the same number of miles on their respective journeys. Show that she is wrong.

....

....

(3 marks)

(b) What is the difference between the distances driven by Niall and Robbie?

....

(3 marks)

Ratio 2

 4 Brass is composed of copper and zinc in the ratio 13 : 7, respectively. A mass of 4 kg of brass is to be made.

(a) Work out the total mass of zinc needed.

Total number of parts of copper and zinc = ... + ...

1 part zinc = 4 ÷ .... = ... kg

7 parts zinc = ... × 7

.... kg of zinc are needed.

(2 marks)

(b) If another 126 g of zinc was added, how much more copper would be needed?

....  How many parts of the ratio does 126 g represent? **(2 marks)**

 5 A 500 g loaf of bread is made with a certain ratio, by mass, of flour to water. There is $1\frac{2}{3}$ more flour than water.

 You will need to use problem-solving skills throughout your exam – **be prepared!**

(a) What is the ratio of flour : water in its simplest form?

Flour : water = $1\frac{2}{3}$: 1

Express $1\frac{2}{3}$ as an improper fraction and then multiply both sides of this ratio by a factor that leaves whole numbers and no fractions.

Ratio is $\dfrac{5}{\text{...........}}$: ....

Multiply by .... to get:

....

(2 marks)

(b) What is the mass of flour used? Give your answer in grams.

....

(1 mark)

 6 Ben and Clare are left £60 000 in a will in the ratio Ben : Clare = 3 : 5. Clare spends £8600 of her share on home improvements. She divides up the remainder in the following ratio: investment : son : daughter = 2 : 1 : 1. Work out the difference between what Ben inherited and what Clare puts into her investment.

£ ....

(4 marks)

 7 The ratio A : B : C, representing goals scored by three football teams, was 9 : 4 : 5. The total number of goals scored by team B was 36. How many goals were scored in total?

....

(3 marks)

 8 The ratio of boys to girls from two classes is 17 : 16. The classes are maths and art. There are 9 girls in the maths class which has a ratio of girls : boys of 3 : 2. If the ratio of girls : boys in the art class is 4 : 5, how many boys take art?

....

(2 marks)

Metric units

1 Convert these quantities.

    (a) 35 mm to cm

        35 ÷ .... = ... cm

> There are 10 mm in 1 cm, so to convert mm to cm, you must divide the number of mm by 10.

        35 mm = ... cm

                                            (1 mark)

    (b) 83 mm to cm

        .... cm   **(1 mark)**

    (c) 42 cm to mm

> If you divide by 10 to convert mm to cm, what do you need to do to convert cm to mm?

        .... mm   **(1 mark)**

    (d) 2300 g to kg

> There are 1000 g in 1 kg, so to convert g to kg you must divide the number of g by 1000.

        2300 ÷ .... = ... kg

        2300 g = ... kg   **(1 mark)**

    (e) 2.35 kg to g

> If you divide by 1000 to convert g to kg, what do you need to do to convert kg to g?

        .... g   **(1 mark)**

    (f) 3450 m to km

        .... km   **(1 mark)**

2 Convert

    (a) 17 cm to mm   (b) 185 mm to cm

    .... **(1 mark)**   .... **(1 mark)**

    (c) 3750 g to kg   (d) 11.1 kg to g

    .... **(1 mark)**   .... **(1 mark)**

    (e) 2.3 litres to ml   (f) 755 ml to litres

    .... **(1 mark)**   .... **(1 mark)**

3 How many 334 ml glasses of water will fill a 2 litre bottle?

> k (kilo) is used to scale up by 1000. 1 kg = 1000 g. m (milli) is used to scale down by 1000. 1 mm = 1 m ÷ 1000. There are 1000 ml in a litre.

    ....   **(2 marks)**

4 In a marathon, there is a water station every 950 m. A marathon is just short of 42 km. How many water stations will be needed?

PROBLEM SOLVED!

> Work out the biggest multiple of 950 m that is smaller than 42 000 m.

> You will need to use problem-solving skills throughout your exam – **be prepared!**

    ....   **(2 marks)**

5 The ratio of butter : flour : sugar = 2 : 3 : 1 is what cooks use when baking biscuits. If a cook uses 1 kg of flour to produce 140 biscuits, how many grams of sugar are in each biscuit?

    ....   **(3 marks)**

Had a go ☐ **Nearly there** ☐ **Nailed it!** ☐

Reverse percentages

1 A piano was bought for £4590 after a discount of 15%. What was its original price?

$100\% - 15\% = 85\% = 0.85$

Original cost of piano $= \dfrac{4590}{85} \times 100$ or $\dfrac{4590 \times 100}{85}$

$= £$............................ **(3 marks)**

> • Take the original price as 100%.
> • £4590 is (100 – 15)% or 85% of the original cost. So the multiplier for the original amount was 0.85.
> • Divide 4590 by 85 to get 1% of the original price.
> • Multiply this by 100 to get the original price.
> Alternatively, you can just divide by 0.85.

2 Sadiq had a 15% pay rise which took his salary to £25 001. What was his original salary?

$100\% + 15\% = 115\% = 1.15$

Original salary $= \dfrac{25\,001}{115} \times 100$

or $\dfrac{25\,001 \times 100}{115}$

$= £$............................

> • Take the original price as 100%.
> • An increase of 15% means (100 + 15)% = 115%. So the multiplier for the original amount was 1.15.
> • Divide £25 001 by 115 to get 1% of the original salary.
> • Multiply this by 100 to get the original salary.
> Alternatively, you can divide the new salary by 1.15.

(3 marks)

3 A house bought last year has increased in value by 8% and is now valued at £253 800. How much was the original purchase price of the house?

$100\% + 8\% = 108\% = 1.08$

$\dfrac{\text{...............} \times 100}{\text{.................}} = £$............................

> Remember, your answer should be less than £253 800.

> Divide current house value into 108 equal parts, then multiply by 100.

Alternatively, $\dfrac{\text{............}}{\text{............}} = £$............................ **(3 marks)**

4 A guitar in a London shop is priced at £840. Pete says that he saw the same guitar for £85 less than that in Lincoln, which was 15% cheaper. Is he correct? Show your working.

.... **(3 marks)**

5 Danni and Han both bought cars last year. The table shows their values now, after a percentage decrease.

Guided

	Current value	Percentage decrease
Danni	£11 780	5%
Han	£11 750	6%

Work out

(a) who bought the more expensive car last year.

.... **(3 marks)**

(b) the difference between the original prices of both cars.

.... **(1 mark)**

Growth and decay

1 If you invested £500 at 10% compound interest, how much should you expect to have after 3 years?

> An increase of 10% means after one year you have 110% of the original sum. To find this multiply the original sum by 1.1. Notice that the power you use with the multiplier represents the number of years in the problem.

Amount at the end of 3 years = £500 × 1.1³ = ...   **(2 marks)**

2 If you invest £4500 for x years at 3% compound interest you will have grown this sum to £5216.73 (to 2 d.p.). Find the value of x, the number of years that the money should be invested for.

100% + 3% = 103% = 1.03

> Work out the multiplier.

After x years the amount will be $(1.03)^x$ × £4500

> Now, find the value of x which gives £5216.73
> Estimate a number for x and substitute it into 1.03^x.
> Keep refining your estimation until you get to the required amount.

x = ...   **(2 marks)**

3 Richard buys a van for £20 000. His financial adviser says that model tends to depreciate in value by 11% each year. If she is right, how much will the van be worth in 4 years' time? (Give your answer to 4 significant figures.)

> A depreciation of 11% means the following year the van will be worth 100% − 11% = 89% of the price. Each year, it is worth 89% of the year before.

Depreciation over 4 years = £20 000 × (0.89)$^{......}$ = ...   **(2 marks)**

4 A formula given by a bank for a saving scheme is $V = 2400 × (1.015)^x$.

(a) What is the interest rate for this saving scheme? Give your answer as a percentage.

....   **(1 mark)**

(b) How much would the investment be worth after 8 years?

....   **(2 marks)**

5 Tessa buys a cast iron printing press for her studio for £3000. It is an antique and is likely to increase in value by 20% per year. Duncan says it will have doubled in value after 5 years because 5 × 20% = 100%.
Is Duncan correct? If not, how much will the printing press really be worth after 5 years?

....   **(2 marks)**

Speed

1 A train takes 4 seconds to travel 120 metres. Work out the average speed of the train in m/s.

$$s = \frac{\ldots\ldots}{4} = \ldots\ldots\ldots\ldots\ldots\ldots\ldots\ldots\ldots$$

Speed of train = ... m/s

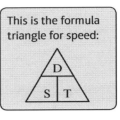

This is the formula triangle for speed:

(2 marks)

2 The distance from Milan to Munich is 495 km. The train between the two cities takes 7 hours 30 minutes. Work out the average speed of the train in km/h.

$$s = \frac{d}{t} = \frac{495}{\ldots\ldots} = \ldots\ldots\ldots\ldots\ldots \text{ km/h}$$

Remember that on a calculator 7 hours 30 minutes is 7.5 hours not 7.3 hours.

(2 marks)

3 Marvin takes 6 hours to get to his destination at an average speed of 72 km/h. How far does he travel?

Rearrange the formula to make *d* the subject.

$$d = \ldots\ldots\ldots \times \ldots\ldots\ldots$$

$$d = \ldots\ldots\ldots\ldots \text{ km}$$

(2 marks)

 Guided

PROBLEM SOLVED!

4 The speed limit in the UK is 70 mph. The distance between Portsmouth and Petersfield is 17 miles. Florence did the journey in her car and it took her 13 minutes. By how much did she break the speed limit?

 You will need to use problem-solving skills throughout your exam – **be prepared!**

Speed limit broken by .... km/h

(2 marks)

 Guided

PROBLEM SOLVED!

5 The distance between Bristol and Penzance is 207 miles and the train between them takes 4 hrs 40 mins. The distance between Birmingham and Norwich is 160 miles and the train journey takes 3 hrs 41 mins. Work out which journey is faster. What is the average speed of the faster journey?

 You will need to use problem-solving skills throughout your exam – **be prepared!**

.... to .... = ... mph

(3 marks)

Density

 1 Work out the density of a piece of plexi-glass with a mass of 18 g and a volume of 10 500 mm³.

$$d = \frac{18}{\text{......}}$$

Density = ... g/mm³

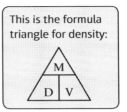
This is the formula triangle for density:

(2 marks)

 2 The density of gold is 19.32 g/cm³. If the mass of a gold bracelet is 80 g, what is its volume?

Rearrange the formula to make *v* the subject.

$$v = \frac{\text{.........}}{19.32}$$

Volume = ... cm³

(2 marks)

 3 Honey has a density of 1.42 g/cm³. The volume of honey in a container is 1250 cm³. Show that there is more than 1 kg of honey in the container.

Rearrange the formula to make *m* the subject.

m = 1.42 × ...... = ... g

Mass of honey in container = ... g = ... kg

(3 marks)

 4 Glass has a density of 400 kg/m³. What would be the mass of a pane of glass measuring 15 cm × 9 cm × 1 cm? Give your answer in grams.

Mass = ... g

(3 marks)

  5 The mass of a stainless steel circular disc, 2 cm thick with a volume of 40π cm³, is 310π grams.

Volume of a circular disc = πr²h

(a) Work out the diameter of the disc to 3 s.f.

$$40\pi = \pi \times \text{......}^2 \times \text{......}$$

$$\frac{40}{\text{......}} = \text{......}^2$$

r = ...

D = ... × ...... = ... cm

You can divide by π on both sides.

D is the diameter, so: 2r = D

(2 marks)

(b) work out the density of stainless steel to 2 significant figures.

$$d = \frac{m}{v} = \frac{\text{......}}{\text{......}}$$

Density = ... g/cm³

This *d* is not the same unknown as the *D* you found in part (a).

(2 marks)

65

Other compound measures

1 A crate exerts a force of 700 Newtons (N) on the floor of a factory. The area of the base of the crate is 3.5 m². Work out the pressure of the crate on the floor.

$P = \dfrac{\text{......}}{\text{......}}$

$P = $ .... N/m²

> Simplify the fraction.

> This is the formula triangle for pressure:
>

(2 marks)

2 A lorry has 6 tyres. The area of each tyre that is in contact with the road is 70 cm². The pressure that the truck exerts on the road is 510 000 N/m². Calculate the total force that the truck exerts on the road.

Total contact area = 6 × .... = ... cm²

$A = $ .... m²

$F = $ .... × .... = ... N

> Be careful with converting the units. There are **not** 100 cm² in 1m²!

> Substitute values into the formula triangle.

(3 marks)

3 A fish tank has the following measurements.

The tank was half filled with water at a rate of 750 ml/sec. How long did it take to get this much water into the tank?

> Remember that to work out the volume, you will need all the dimensions in the same units.

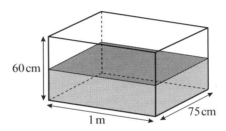

$V = $ .... × .... × .... = ... cm³

Amount of water in tank = ... × .... = ... cm³

Time taken $= \dfrac{\text{......}}{\text{......}} = $ .... seconds = ... minutes

> Remember that 1 cm³ is equal to 1 ml.
> $t = \dfrac{V}{R}$

(3 marks)

4 A guitar exerts a force of 80 N on a guitar stand. The surface area of the guitar stand that the guitar rests on is 48 cm². What is the pressure exerted on the stand?

$P = $ ....

> Remember to give units of measurement in your answer.

(3 marks)

5 A 2400 litre vat of orange juice has filled 3200 bottles in 20 minutes.

(a) How much orange juice is in one bottle?

....

(1 mark)

(b) How long did it take to fill each bottle?

....

(1 mark)

Proportion

1 A bus company sells 3 monthly bus tickets on Friday, for £105 in total. The next Monday the company sells 8 of these tickets. Work out how much the company receives for the 8 tickets sold on Monday.

£105 ÷ 3 = ...

.... × 8

....

> Find the cost of one ticket.

> Multiply the cost of one ticket by the number of tickets needed.

(2 marks)

2 Thelma bought 6 crusty rolls for £1.98. How much would it cost to buy 25 crusty rolls?

....

(2 marks)

3 One vat of vegetable oil fills 15 bottles that each hold 1.5 litres. How many vats of vegetable oil will be needed to fill 315 bottles?

....

(2 marks)

4 A machine makes 250 metal discs in 1 hour. How many discs can three of these machines make in 6 minutes?

> Use the fact that 6 minutes = $\frac{1}{10}$ of an hour to work out how many metal discs one machine makes in 6 minutes. Then multiply this by 3 for the 3 machines.

....

(2 marks)

5 It takes 3 people 4 days to paint 5 offices in a block of 30 offices. How long would it take 12 people to paint every office in the block?

> You will need to use problem-solving skills throughout your exam – **be prepared!**

> Work out how many days three people would take to paint the whole block.

3 people would take 4 × $\frac{30}{............}$ = ... days.

6 people would take .... days.

12 people would take .... days.

> If you double the amount of people it will take half the time.

(3 marks)

6 A lottery jackpot was shared among 30 people. Each person received £350. The following week a group of 7 people won the same-sized jackpot. Calculate how much each person in this second group received.

Total lottery jackpot = ... × .... = ...

Each person in the second group received .... ÷ 7 = ...

£ ....

(2 marks)

7 A 900 g bag of tomatoes costs £4.80 and a 1.2 kg bag costs £6.42. Show that the 900 g bag is better value for money.

(3 marks)

Proportion and graphs

1 The acceleration of a mass is directly proportional to the force on the same mass.
If $a = 38$ m/s and $F = 720$ N, work out the value of F when $a = 66.5$. m/s.

> Set the ratios as equal fractions. Then rearrange to solve the equation in F
> $\frac{720}{38} = \frac{F}{66.5}$ You could also have compared the values in the form $F = ka$: $F = \frac{720}{38} \times 66.5$

$F = \dfrac{720 \times \text{............}}{\text{..............}} = \text{.......................}$ N   **(2 marks)**

2 The electrical resistance (R), in ohms, of a piece of copper wire is proportional to its length. A length (L) of 12 m has a resistance of 30 ohms.
How long would a piece of this wire be if it had a resistance of 50 ohms?

$\dfrac{30}{12} = \dfrac{50}{L}$

> Set the ratios as equal fractions, rearrange and solve.

$L = \dfrac{50 \times \text{............}}{\text{..............}} = \text{.......................}$ m   **(2 marks)**

3 The exchange rate of euros to pounds is €1.17 to £1.

If $x = £$ and $y = €$, write an equation for the exchange rate, including the constant of proportion.

$y = \text{........................}$ **(2 marks)**

4 The graph shows the variation with force (F) of the extension (s) of a steel spring.

(a) What would the force (F) be if the spring was extended to 10 cm?

> Use the graph to work out the force (F) when the spring is 5 cm long and then double this figure.

....   **(2 marks)**

(b) If a force of 40 N was exerted on the spring what would its extension measure?

> How many times bigger is 40 N than 10 N?

....   **(2 marks)**

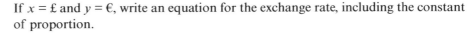

5 The graph shows the relationship between pressure in pascals, and volume in litres.

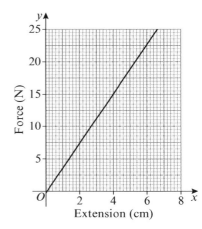

(a) Prove, by taking two coordinates, the equation of proportionality $V = \dfrac{k}{P}$ and then work out the value of k.

.... **(3 marks)**

(b) Use the equation to solve the pressure needed to maintain a volume of 5 litres.

.... **(2 marks)**

Problem-solving practice 1

1 In his recent exams, Jack scored $\frac{43}{50}$ in mathematics, 87% in French and $\frac{51}{60}$ in art.

Order the subjects by score from best to worst.

....

(2 marks)

2 In a box of 140 apples for sale, the grocer found 17 of them were bruised and could not be sold. She said that there were very close to 83% left for sale. Do you agree? Show your working.

....

(3 marks)

3 The same pair of shoes is sold by different companies online.

Decide whether Shoe online or E shoe shop is better value for money. You must show your working.

Shoe online
£65
+ 20% VAT

Free delivery

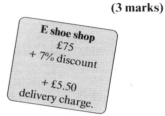

E shoe shop
£75
+ 7% discount

+ £5.50 delivery charge.

....

(3 marks)

4 In a recipe for a fruit cake, it says you need 225 g of flour and 110 g of sugar. Angelica has only 210 g of flour. How much sugar will she need to make a cake with this amount of flour? Give your answer to the nearest gram.

....

(4 marks)

5 Some cubes measuring 2 cm × 2 cm × 2 cm are made from iron or tin or brass for use in a design studio. The density of each metal is shown in the table.

Metal	Density g/cm³
Iron	7.9
Tin	7.3
Brass	8.4

(a) A designer uses one cube of each metal to make a model. What is the mass of metal in the model?

.... g

(2 marks)

(b) Serena needs to make a model that weighs 184 g exactly. Which combination of three cubes would give this total mass?

....

(2 marks)

69

Problem-solving practice 2

6 Five machines produce 27 000 litres of orange juice in 8 hours. How much orange juice would 4 machines produce in 9 hours? You will need to show your working.

> First work out, by dividing, how much orange juice 1 machine produces in 1 hour. Then use multiplying to find out the required information.

....

(3 marks)

7 A 3-box pack of 'Wheatybangs' costs £3.55, a 5-box pack costs £5.90 and an individual box costs £1.19. Which of the three ways of buying 'Wheatybangs' is the best value for money? Show your working.

> Find how much one 'Wheatybang' box costs in the 3 pack and the 5 pack and then compare these values with the cost of 1 individual box.

....

(3 marks)

8 The recipe for 20 doughnuts is given. Sasha has made a list of the ingredients available at home. What is the maximum number of doughnuts that she can make with the ingredients she has?

Guided

Recipe for 20 doughnuts	Ingredients available
500 g flour	1200 g flour
60 g caster suger	65 g caster suger
15 g yeast	18 g yeast
125 g butter	145 g butter
4 eggs	7 eggs

.... doughnuts

(3 marks)

9 Nedra and Hans are saving for a house deposit of £35 000. Nedra invests £15 500 in an ISA for 3 years at a compound interest rate of 5.5%. Hans invests £15 200 for the same period of time in a different ISA at a compound interest rate of 5.75%.

Guided

(a) Which of them will have gained more money at the end of the 3 years? You must show your working.

....

(3 marks)

(b) Which of them will have more money in total?

....

(1 mark)

(c) If Nedra and Hans combine their savings at the end of the 3 years, will they have enough for a deposit? You must show your working.

....

(3 marks)

Symmetry

1 Draw the lines of reflective symmetry in the letters shown.

| M | | O |

> Reflective symmetry means that one side of the line is a mirror image of the other side. Often there is more than one line of symmetry.

(2 marks)

2 Rashid says, 'A polygon has the same number of lines of symmetry as number of sides.' Is Rashid's statement true? If not, how would you change it?

.... **(2 marks)**

3 In the diagrams below, shade one extra square or triangle to give a shape with

(a) 1 line of symmetry.   (b) order 2 rotational symmetry.

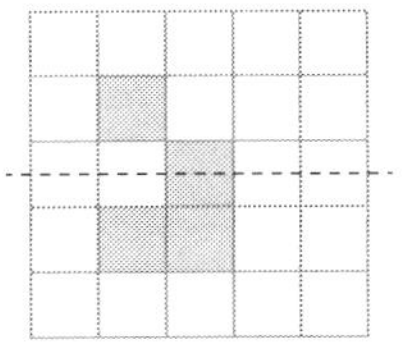

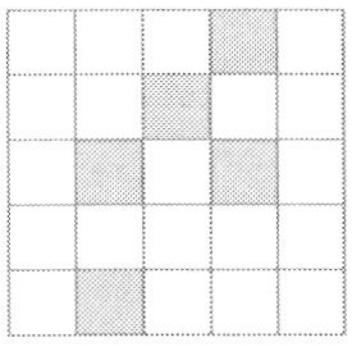

> To test for rotational symmetry, imagine putting a pin in the middle of the image and twirling it round. How many times would it look the same in one revolution?

(c) 2 lines of symmetry.

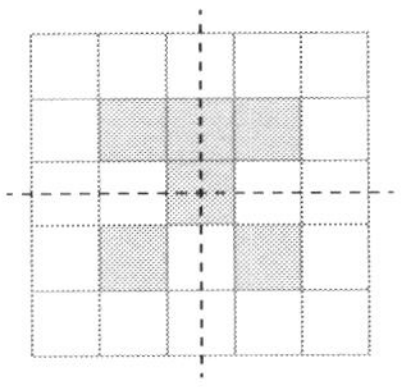

(d) order 3 rotational symmetry.

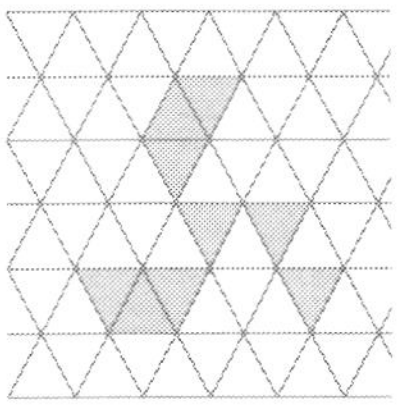

(4 marks)

4 The diagrams show a regular octagon and an irregular octagon.

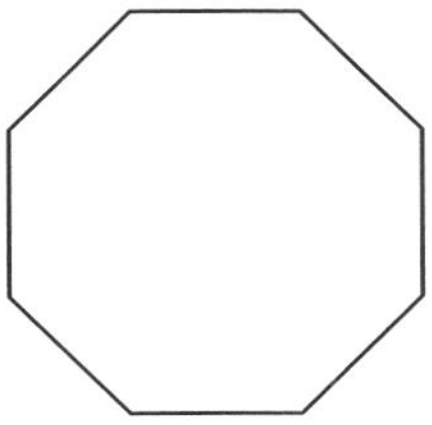

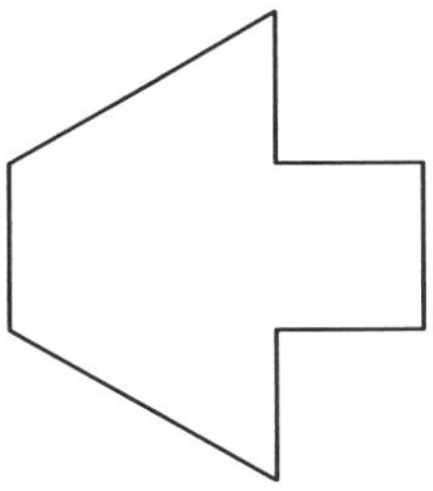

(a) Draw one line of symmetry on each octagon.

(1 mark)

(b) How many possible lines of symmetry are there on the irregular octagon?

.... **(1 mark)**

(c) Decide the order of rotational symmetry for the

(i) regular octagon.

....

(1 mark)

(ii) irregular octagon.

....

(1 mark)

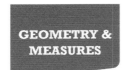

Quadrilaterals

1   Label all the angles on each quadrilateral below with letters *a*, *b*, or *c*. Within each shape, if angles are equal the letter needs to be the same.

> Some shapes have lines of symmetry shown, and some angles are already marked.

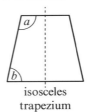
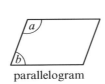

rectangle   isosceles   parallelogram
                 trapezium

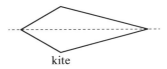

square   rhombus   kite

(6 marks)

2   Draw quadrilaterals with the following properties. Label each one with its name.

> The shape below has two different side lengths.

(a) One pair of parallel sides and one line of symmetry.

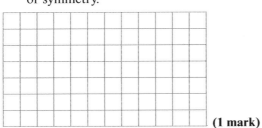

(1 mark)

(b) Two pairs of parallel sides with unequal diagonals.

(1 mark)

(c) Only two equal obtuse angles.

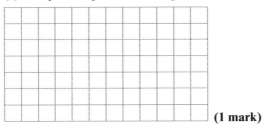

(1 mark)

(d) Two pairs of equal angles and no rotational symmetry.

(1 mark)

(e) Order 4 rotational symmetry.

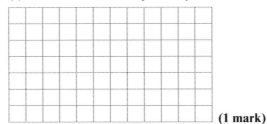

(1 mark)

(f) Only two equal acute angles.

(1 mark)

Angles 1

1 Name the type of angle represented by the following.

> An angle less than 90° is acute.
> An angle between 90° and 180° is obtuse.
> An angle greater than 180° is reflex.

(a) 167°

obtuse .... **(1 mark)**

(b) 52°

.... **(1 mark)**

(c) 203°

.... **(1 mark)**

(d) 90°

.... **(1 mark)**

> Remember: this angle has a special name.

2

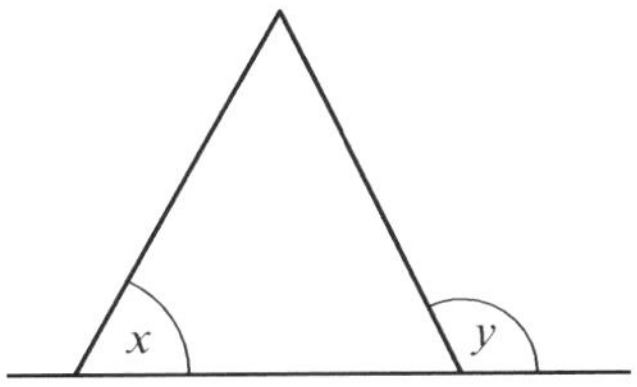

(a) What type of angle is x?

....

Explain your answer.

.... **(1 mark)**

> Think about whether the angles are bigger or smaller than 90°.

(1 mark)

(b) What type of angle is y?

....

(1 mark)

Explain your answer.

.... **(1 mark)**

3 (a) What is the value of angle x?

.... **(1 mark)**

(b) What type of angle is angle x?

....

91° x

(1 mark)

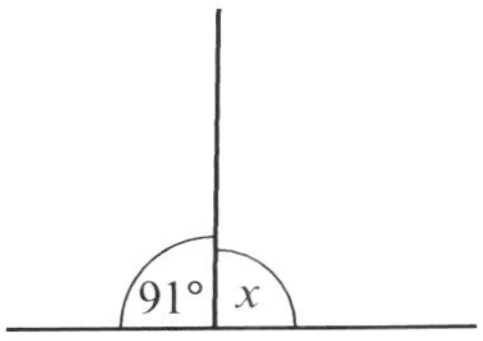

4 In this diagram, angle $a = 4y°$ and angle $b = 5y°$.

(a) Calculate the value of y.

....

a
b

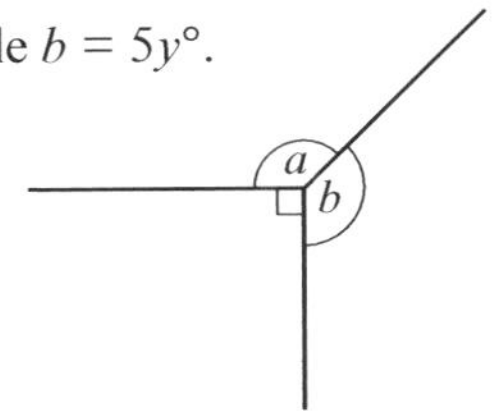

> The sum of the angles around a point is 360°. Notice you are also given the third angle.

(2 marks)

(b) Write down the size of angle a.

..............................°

(1 mark)

(c) Write down the size of angle b.

..............................°

(1 mark)

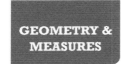
Had a go ☐ **Nearly there** ☐ **Nailed it!** ☐

Angles 2

5 Work out the value of angle x.

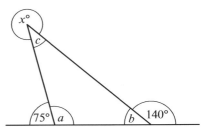

$180° - 75° = a$

$a = \ldots\ldots\ldots\ldots$

$180° - 140° = b$

$b = \ldots\ldots\ldots\ldots$

$180° - (a + b) = c$

$c = \ldots\ldots\ldots\ldots$

$x = 360° - c$

$x = \ldots\ldots\ldots\ldots$

> Angles on a straight line add up to 180°.

> Angles in a triangle add up to 180°.

> Angles around a point add up to 360°.

(3 marks)

6 One isosceles triangle and two equilateral triangles are placed together.
Work out the value of angle x.

> First, label all the angles inside the equilateral triangles. Then, calculate angle a using the sum of angles around a point.

$a = \ldots\ldots\ldots°$

> Calculate angle b using the formula $2a + b = 180°$.

$b = 180° - \ldots\ldots\ldots° = \ldots\ldots\ldots°$

$x = 360° - \ldots\ldots\ldots° - \ldots\ldots\ldots° = \ldots\ldots\ldots°$

> Calculate angle x using the sum of angles around a point.

(4 marks)

7 A slanting line is drawn through a pair of parallel lines.

Guided

(a) Work out angle a.

$\ldots\ldots\ldots\ldots\ldots\ldots\ldots\ldots\ldots$

(1 mark)

(b) Work out angle b.

$\ldots\ldots\ldots\ldots\ldots\ldots\ldots\ldots\ldots$

(1 mark)

(c) Work out the value of $a + d$.

$\ldots\ldots\ldots\ldots\ldots\ldots\ldots\ldots\ldots$

(1 mark)

(d) If the angle labelled 56° was changed to 43°, would this change the value of $a + d$? Give a reason for your answer.

$\ldots\ldots\ldots\ldots\ldots\ldots\ldots\ldots\ldots\ldots\ldots\ldots\ldots\ldots\ldots\ldots\ldots\ldots\ldots\ldots\ldots\ldots\ldots\ldots\ldots\ldots\ldots\ldots\ldots\ldots\ldots\ldots\ldots$

(1 mark)

Solving angle problems

1 A triangle is drawn between two parallel lines, with the angles as marked.

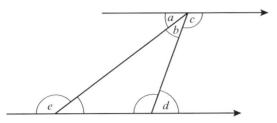

(a) Label each of the unknown angles with an
   expression, using alternate angle facts.

> Angles on a straight
> line add up to 180°.

(2 marks)

(b) Rewrite angle *d* in terms of *a*, *b* or *c*.

....

> Alternate angles are equal.
> Co-interior angles add up to 180°.

(2 marks)

(c) Rewrite angle *e* in terms of *a*, *b* or *c*.

....

(2 marks)

2 Two parallel lines cut a straight line. Work out the value of *x*.

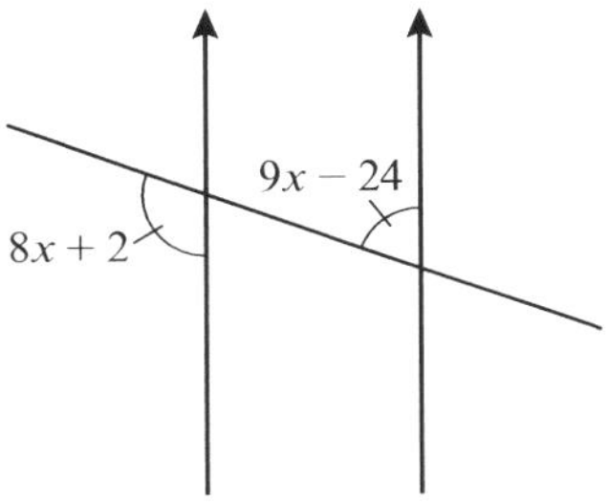

$9x - 24$

$8x + 2$

$(8x + 2) + (9x - 24) =$ ....°

............$x =$ ....

$x =$ ....°

(4 marks)

3 Work out the value of *x*.

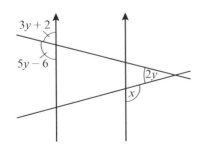

$3y + 2$

$5y - 6$

$2y$

x

....

(4 marks)

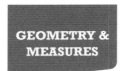

Had a go ☐   **Nearly there** ☐   **Nailed it!** ☐

Angles in polygons

1 The diagrams show two identical regular octagons.

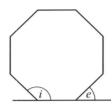 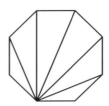

(a) Work out the value of i (the interior angle).

$8 \times i = (8 - 2) \times 180° = $ ....°

So, $i = \dfrac{\dots\dots}{8}$

$i = $ ....°

> The sum of the interior angles of a regular polygon = $(n - 2) \times 180°$ where n is the number of sides.

(2 marks)

(b) Work out the value of e (the exterior angle).

$i + e = 180°$

So, $e = 180° - $ ....°

$e = $ ....°

> The interior angle and its corresponding exterior angle always add up to 180°.

(2 marks)

(c) Erica gives a method for finding the size of the interior angles: 'multiply the number of triangles in the diagram by 180° and then divide by 8'. Prove that this method is correct.

> Compare Erica's way of calculating i with the method used in (a). Are the answers the same?

....

.... **(3 marks)**

2 A regular polygon has interior angles of 156°.

(a) Calculate the exterior angle of this polygon.

....

> Angles on a straight line add up to 180°.

(1 mark)

(b) How many sides does the polygon have?

....

> The sum of exterior angles of any polygon is 360°. Divide this by the exterior angle to find out how many angles (and sides) the regular polygon has.

(2 marks)

3 The diagram shows part of a regular 11-sided polygon. Work out the value of x to 4 s.f.

$x = $ ....

(2 marks)

4 Work out the value of n in the regular pentagon.

> You will need to use problem-solving skills throughout your exam – **be prepared!**

> Use what you know about isosceles triangles and the sum of external angles. Find an isosceles triangle that might help to label some useful angles.

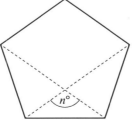

$n = $ ....

(4 marks)

Time and timetables

1 Write the following 24-hour clock times in 12-hour form.

(a) 15:45

.... .45 pm   **(1 mark)**

(b) 18:05

.... .05 pm   **(1 mark)**

(c) 10:09

.... am   **(1 mark)**

(d) 09:59

....   **(1 mark)**

2 Write the following times using the 24-hour clock.

(a) Quarter past seven in the evening.

19 ....   **(1 mark)**

(b) Thirteen minutes to eleven at night.

.... 47   **(1 mark)**

(c) Three minutes past four in the morning.

0............   **(1 mark)**

(d) Eighteen minutes to two in the morning.

....   **(1 mark)**

3 A coach left Calais at 07.15 and arrived in Rome the next day at 17.55. How long did the journey take?

.......hours ....

0715   0715   1755

```
A number line might be useful!
```

....   **(2 marks)**

4 A bus leaves Leeds at 08.07 and takes 85 minutes to get to Halifax via Bradford.
The bus then returns direct to Leeds, taking 37 minutes.
At what time does the bus arrive back in Leeds?

....   **(2 marks)**

5 Below is an extract from a railway timetable, with some information missing.

Departs	Waterloo	15.28	16.06	16.28	....
Arrives	Havant	16.53	....	17.51	18.28

(a) If you depart from Waterloo at 15.28, how long is this journey to Havant in hours and minutes?

....   **(1 mark)**

(b) The 16.06 departure from Waterloo takes 1 hour and 38 minutes. Calculate its arrival time in Havant.

....   **(1 mark)**

(c) The service that arrives in Havant at 18.28 takes 1 hour and 29 minutes. Calculate its departure time from Waterloo.

....   **(1 mark)**

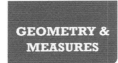
Had a go ☐   Nearly there ☐   Nailed it! ☐

Reading scales

1 Write the value shown on the number line.

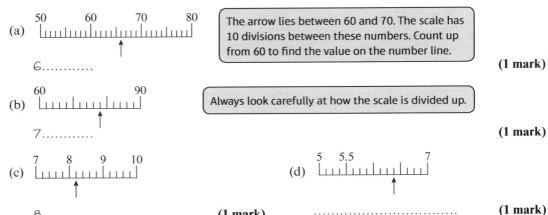

(a)

> The arrow lies between 60 and 70. The scale has 10 divisions between these numbers. Count up from 60 to find the value on the number line.

6...........

(1 mark)

(b)

> Always look carefully at how the scale is divided up.

7...........

(1 mark)

(c)

8...........

(1 mark)

(d)

....

(1 mark)

2 Mark the following numbers on the number line with an arrow.

(a) 131

(1 mark)

(b) 18.8

(1 mark)

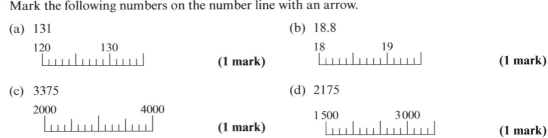

(c) 3375

(1 mark)

(d) 2175

(1 mark)

3 Write down the mass shown.

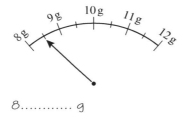

8........... g

(1 mark)

4 Here are some weighing scales. Show where the arrows would point for the following masses.

(a) 8.3 g

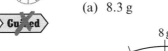

(1 mark)

(b) 43 g

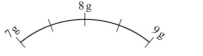

(1 mark)

5 Two households on an island use generators for their electricity. The fuel tank of the Davis family's generator has a capacity of 42 litres, and the fuel tank of the Llwellyn family's generator has a capacity of 38 litres.

Which fuel tank currently contains more fuel and how much does it have?

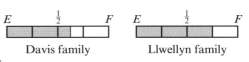

Davis family   Llwellyn family

....

....

> You will need to use problem-solving skills throughout your exam – **be prepared!**

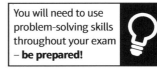

(3 marks)

Perimeter and area

1 The diagram shows a 6 cm × 6 cm grid.

(a) Work out the perimeter of the shaded shape.

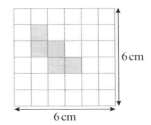

> Start at a corner and move around the outside until you arrive back to where you started. Count each side of the shaded squares along the way.

Perimeter = ... cm   **(1 mark)**

(b) Work out the area of the shaded shape.

> Count all the shaded squares. Alternatively, break the shape up into smaller rectangles, then sum their areas.

Area = ... cm²   **(1 mark)**

2 The diagram shows another 6 cm × 6 cm grid.

(a) Work out the area of the shaded shape.

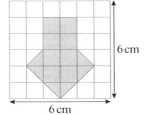

> Where squares are split on the diagonal you can count up in half squares.

Area = ... cm²   **(2 marks)**

(b) Work out the fraction of the grid that is shaded.

$$\frac{\text{Number of squares shaded}}{\text{Number of squares in grid}} =$$

....   **(1 mark)**

3 Work out the perimeter of

(a) this rectangle.

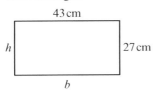

> A useful formula for finding the perimeter (*P*) of a rectangle is *P* = 2(*b* + *h*)

P = ... cm   **(1 mark)**

(b) this regular pentagon.

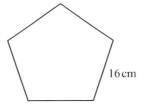

> There are five sides all of the same length.

P = ... cm   **(1 mark)**

4 The irregular octagon has a line of reflective symmetry.

(a) Work out its perimeter.

P = ... cm   **(2 marks)**

(b) Work out its area.

A = ... cm²   **(3 marks)**

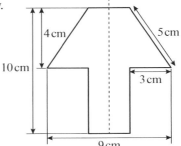

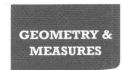

Had a go ☐ Nearly there ☐ Nailed it! ☐

Area formulae

1 Find the areas of the following shapes.

(a)

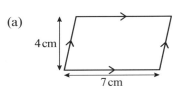

4 cm

7 cm

> Area of a parallelogram is base × height

Area = ... × ....

= .... cm² **(2 marks)**

(b)

5 cm

5 cm

8 cm

> Area of trapezium
> $$= \frac{\text{parallel side } a + \text{parallel side } b}{2} \times \text{height}$$

Area = $\dfrac{\text{.................} + \text{.................}}{2}$ × ....

= .... cm² **(2 marks)**

(c)

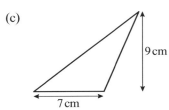

9 cm

7 cm

> Area of a triangle
> $$= \frac{\text{base × height}}{2}$$

= .... cm² **(2 marks)**

(d)

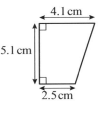

4.1 cm

5.1 cm

2.5 cm

= .... cm² **(2 marks)**

2 A rectangle has a base of 4.2 cm and height h cm.
The ratio of the base to the height is 3 : 2.

h cm

4.2 cm

(a) Calculate the area of the rectangle.

> First find one part of the ratio. h = 2 parts.

= .... cm² **(2 marks)**

(b) A different rectangle with base 4.2 cm and
height g, has an area of 29.4 cm².
Express the ratio of the sides of the rectangle
$b : g$ in its simplest form.

> First work out the value of g to form the ratio. Then multiply both b and g by the same fraction to get whole numbers.

.... **(3 marks)**

Guided

3 The triangle and trapezium are equal in area. Calculate the length of side x.

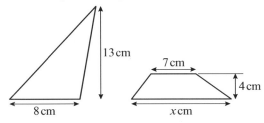

13 cm

8 cm

7 cm

4 cm

x cm

x = ... cm **(4 marks)**

Solving area problems

1 The compound shape has been divided into a triangle and 2 rectangles. Calculate the length of side x if the area of the shape is 186 cm².

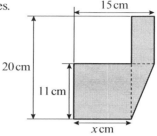

> The compound shape (shaded) lies inside a 15 cm × 20 cm rectangle. Subtract the areas that are not shaded from the area of the rectangle. You will need to write them as algebraic expressions.

Area of outer rectangle = 15 cm × 20 cm = ... cm²

Area of unshaded rectangle = x cm × .... cm = ... cm²

Area of unshaded triangle = $(15 - x) \times$ .... cm = ... cm²

Area of compound shape = outer rectangle − area of unshaded (rectangle + triangle)

= .... − (................. + ...) = ... cm²

$x =$ .... cm

> Solve the linear equation in terms of x.   **(3 marks)**

PROBLEM SOLVED!

2 The area of the triangle is 5 times that of the square. Calculate the length of side x.

> The area of a triangle is found by multiplying the base length by the height, then dividing by 2.

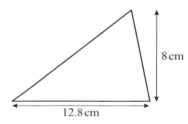

Triangle area = $\frac{1}{2} \times 12.8 \times 8 =$ ....

Divide this by 5: = ...

> You will need to use problem-solving skills throughout your exam – **be prepared!**

$x^2 =$ ....

$x =$ .... cm

(4 marks)

Guided

3 Wallpaper is sold in rolls that measure 40 cm by 4.5 metres. Angharad has to wallpaper a wall with a door and window as shown in the diagram. How many rolls of wallpaper will she need to do this job?

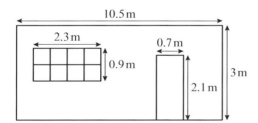

.... rolls

(4 marks)

3D shapes

1 What is the name of a 3D shape with 7 faces, 2 of which are identical but share no edges?

.... **(1 mark)**

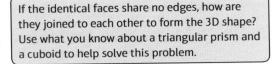

If the identical faces share no edges, how are they joined to each other to form the 3D shape? Use what you know about a triangular prism and a cuboid to help solve this problem.

2 The surface areas of the following prisms are given. Calculate the unknown side length.

(a) Surface area = 96 cm²

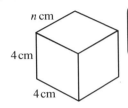

n cm

4 cm

4 cm

Remember, a cuboid or cube is also a rectangular prism.

Surface area = 2(4 × 4) + 4(4 × *n*)

   = 32 + 16*n*

Surface area = 32 + 16*n* = 96

16*n* = ...

n = ... **(2 marks)**

(b) Surface area = 148 cm²

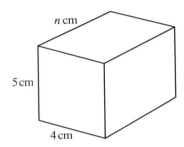

n cm

5 cm

4 cm

Surface area = 148 cm²

   = 2(4 × 5) + 2(5*n*) + 2(4*n*)

148 = 40 + ...*n*

n = ... **(2 marks)**

3 Work out the surface area of the triangular prism shown.

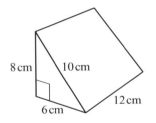

8 cm   10 cm

6 cm   12 cm

Surface area = 2 × area of triangular faces + areas of 3 different rectangles.

= $2(\frac{1}{2} \times 6 \times 8)$ + (10 × 12) + (6 × ....) + (............ × ....)

= .... + ... + ... + ...

= .... cm²   **(2 marks)**

4 The diagram shows a metal L-shaped hexagonal prism bracket.

(a) Calculate the surface area of the prism.

....   **(2 marks)**

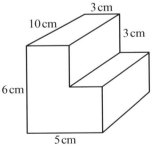

3 cm

10 cm   3 cm

6 cm

5 cm

(b) Ieuan must paint 30 of these brackets. A tin of paint covers 600 cm². How many tins of paint does Ieuan need to do this?

....   **(3 marks)**

Volumes of cuboids

1 What is the volume of each cuboid?

(a)

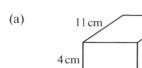

Volume = (7 × 4) × ....

= .... cm³   **(2 marks)**

(b)

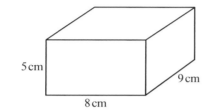

Volume = ... × .... × ....

= .... cm³   **(2 marks)**

(c)

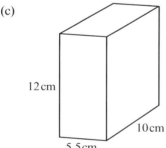

.... cm³   **(2 marks)**

2 A steel girder is in the form of a cuboid with a volume of 10 080 cm³. Work out its length giving your answer in metres.

Let x = the length of the girder

x × 5.5 × 12 = 10 080

$x = \dfrac{.............}{.............} = ...$

....

length

5.5 cm

12 cm

> Call the length x. Volume = 5.5 × 12 × x. You can solve this to find x, the length in cm.

(2 marks)

PROBLEM SOLVED!

3 Container **A** is half full of water. Container **B** is also half full, with the same amount of water. Work out the height of container **B**.

> You will need to use problem-solving skills throughout your exam – **be prepared!**

> Work out the capacity of **A**. Write an expression for the capacity of **B**. Now form an equation in *h* and solve it.

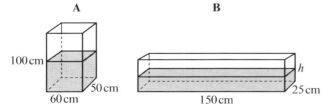

A

100 cm

60 cm   50 cm

B

150 cm   25 cm   h

h = ... cm   **(3 marks)**

Prisms

1 Find the volumes of the following prisms.

(a)

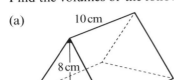

10 cm
8 cm
9 cm

> Find the area of the end faces first then multiply by the length of the prism. Remember: volume is always expressed in 'cubic units'.

Area of triangular faces = $\frac{1}{2}$ × 9 cm × .... cm = ... cm²

Volume = length × area of triangular faces = ... × ....

= .... cm³

(2 marks)

(b)

11 cm
9 cm
5 cm
7 cm

Area of trapezium faces = $\frac{1}{2}(7 + 9) × 5$

= .... cm²

Volume of prism = length × area of

trapezium faces = ... × ....

= .... cm³   **(2 marks)**

(c)

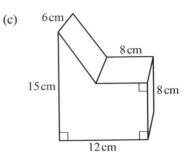

6 cm
8 cm
15 cm
8 cm
12 cm

.... cm³

(2 marks)

2 The net of an isosceles trapezoidal prism is shown.

> Use the dimensions you are given, and the fact that some sides are marked as equal, to find the missing dimensions.

(a) Work out the perimeter of the net.

....

(b) Work out the surface area of the net.

> Calculate the separate areas and add them together.

....

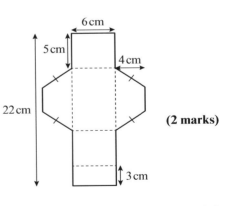

6 cm
5 cm
4 cm
22 cm
3 cm

(2 marks)

(3 marks)

PROBLEM SOLVED!

(c) Calculate the volume of the prism made from this net.

> You need the area of one trapezium, multiplied by the length of the prism.

....

> You will need to use problem-solving skills throughout your exam – **be prepared!**

(3 marks)

Units of area and volume

1 Convert

(a) 4 cm² into mm²

4 cm × 1 cm

= .... mm × .... mm

.... mm²

> An area (cm²) is a multiple of two factors. In this case it could be 4 cm × 1 cm or 2 cm × 2 cm. Choose one, and convert each factor to mm by multiplying by 10.

(2 marks)

(b) 3 km² to m²

3 km² = 3 km × 1 km

= 3000 m × 1000 m

= .... m²

(2 marks)

(c) 800 mm² to cm².

> Choose any two factors that multiply to make the number: 800 = 40 × 20.

800 mm² = 40 mm × 20 mm

= .... cm × .... cm

.... cm²   **(2 marks)**

2 Convert

(a) 4 m³ into cm³

4 m³ = 1 m × 2 m × 2 m

= .... cm × .... cm × .... cm

.... m³

> Cubic measurement indicates volume or a 3D shape. As with area, it is a good idea to find 3 numbers that multiply together to give the number in question.

(2 marks)

(b) 30 cm³ into mm³

30 cm³ = 1 cm × 3 cm × 10 cm

= ...... mm × ...... mm × ...... mm

= .... mm³   **(2 marks)**

(c) 250 mm³ into cm³

250 mm³ = 25 mm × 10 mm × 1 mm

= 2.5 cm × .... cm × .... cm

.... cm³

(2 marks)

3 Convert

(a) 2700 cm³ into litres

.... litres   **(2 marks)**

(b) 9 m³ into litres

.... litres   **(2 marks)**

(c) 17 m³ into litres.

.... litres   **(2 marks)**

4 How many litres of water are needed to fill each container?

(a)

5 cm   12 cm   16 cm

(b)

3 m   10 m   9 m

> You will need to use problem-solving skills throughout your exam – **be prepared!**

.... litres   **(2 marks)**   .... litres   **(2 marks)**

Translations

A vector describes a movement or translation from one position to another. Left or right movement is described by *x*, up and down movement by *y*. Negative *x* = left, negative *y* = down. Vectors are written in the form $\begin{pmatrix} x \\ y \end{pmatrix}$.

1 Write down the following in vector notation.

(a) 4 right, 5 up

$\left(\begin{matrix} \ldots \\ \ldots \end{matrix} \right)$

| Right and up are both positive. |

(1 mark)

(b) 7 right, 3 down

$\left(\begin{matrix} \ldots \\ \ldots \end{matrix} \right)$

(1 mark)

(c) 1 left, 3 up

$\left(\begin{matrix} \ldots \\ \ldots \end{matrix} \right)$

(1 mark)

(d) 4 left

$\left(\begin{matrix} \ldots \\ \ldots \end{matrix} \right)$

(1 mark)

2 Translate

A translation moves each point by the same amount, keeping the shape the same.

(a) shape **B** through vector $\begin{pmatrix} 5 \\ 2 \end{pmatrix}$.

Move each of the three corners of triangle **B** by 5 places to the right and by 2 places up.

When you have translated a shape using a vector, you can say you have **mapped** the shape to a new position.

(2 marks)

PROBLEM SOLVED!

(b) shape **A** through vector $\begin{pmatrix} 6 \\ -8 \end{pmatrix}$.

Move each of the three corners of triangle **A** by 6 places to the right and by 8 places down.

(2 marks)

You will need to use problem-solving skills throughout your exam – **be prepared!**

3 (a) Describe the single transformation in vector form that maps **P** to **R**.

....

.... **(2 marks)**

(b) Describe the single transformation in vector form that maps **Q** to **P**.

....

.... **(2 marks)**

Had a go ☐   Nearly there ☐   Nailed it! ☐

Reflections

1 The shapes are reflected in the dashed line. Draw their reflections.

(a)

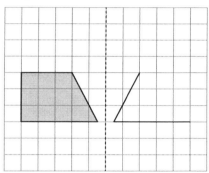

(2 marks)

(b)

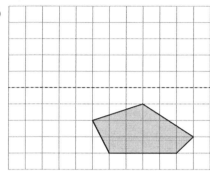

(2 marks)

(c)

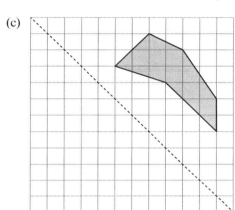

(2 marks)

> When reflecting in a diagonal line, count the number of diagonals of the squares between each vertex of the shape and the reflection line. Count the same number of diagonals on the other side of the line to draw the reflection.

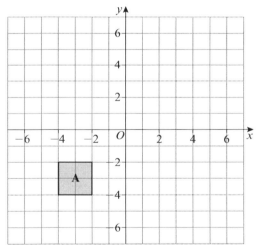

2 Shape **A** is plotted on a grid.

(a) Reflect shape **A** in the line $y = 0$.

> Count the vertical and horizontal distance from each corner to the line and move it the same amount the other side of the line.

(2 marks)

(b) Reflect shape **A** in the line $x = -1$.

(2 marks)

3 (a) Describe the single transformation that maps **P** to **Q**.

   ....

   (2 marks)

   (b) Describe the single transformation that maps **Q** to **R**.

   ....

   (3 marks)

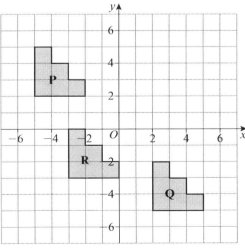

Rotations

Rotations have a centre of rotation (a point), a direction (clockwise or anticlockwise), and an angle (e.g. 90°).

1 The grid shows a trapezium (**T**) and a triangle (**TR**).

(a) Rotate the trapezium 90° clockwise through the point (−2, 1). Label the image **T′**.

(2 marks)

(b) Rotate the triangle 180° through the point (−1, −1). Label the image **TR′**.

(2 marks)

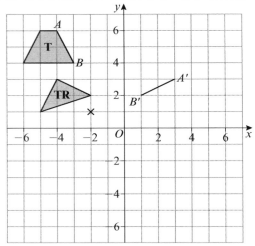

2 The grid shows shape **A**.

(a) Rotate shape **A** 90° clockwise about the point (1, 2). Label this image **A′**.

> Place a piece of tracing paper over the grid and fix it at the point of rotation using the end of a pair of compasses. Mark the vertices on the tracing paper and then rotate it 90° to find the rotated vertices.

 You will need to use problem-solving skills throughout your exam – **be prepared!**

(2 marks)

(b) Rotate shape **A** 180° about the point (1, 0). Label this image **A′**.

(2 marks)

PROBLEM SOLVED!

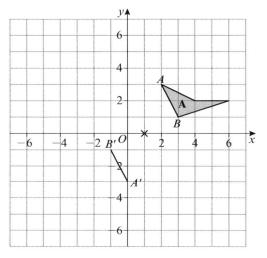

3 Three identical triangles are shown on the grid.

Identical shapes are known as congruent.

 Guided

(a) Describe the single transformation that maps shape **A** to shape **B**.

....

....

(3 marks)

(b) Describe the single transformation that maps shape **B** to shape **C**.

....

....

(3 marks)

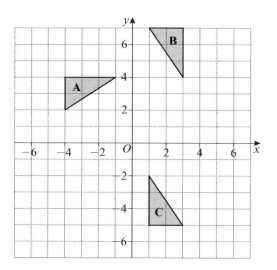

88

Enlargem

1   Work out the scale factor enlargement of rectangle **A**
    to rectangle **B**.

> Find a side length on **A** and compare it with the
> corresponding side on **B**.
> Find the multiplier that transforms the length on **A** to
> the length on **B**. Check this is the same for all other
> corresponding pairs of lengths.

.... **(2 marks)**

2   Here are two similar trapeziums.

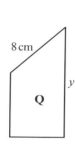 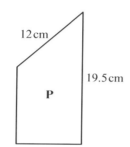

> Similar shapes have exactly the
> shape but they are different sizes.

> Divide the ratio 12 : 8 to get a unit ratio in
> the form 1: *n*. The scale factor is equal to *n*.

(a) Work out the scale factor enlargement of shape **P** to shape **Q**.

.... (2 marks)

(b) Calculate the length of side *y*.

$$\frac{12}{8} = \frac{19.5}{y}$$

$12y = $ .... × .... = ...

$y = \frac{\text{......}}{\text{......}} = $ ....   **(2 marks)**

3   Describe fully the single transformation that maps shape **J** to **K**.

This is an ....

The centre of enlargement is (............, ...)

> Compare corresponding
> sides or lengths to work
> out the scale factor.

> You will need to use
> problem-solving skills
> throughout your exam
> – **be prepared!**

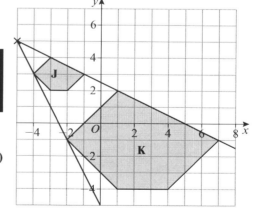

Ratio is 1: ....

Scale factor is ....   **(3 marks)**

Pythagoras' theorem

goras' theorem states that, in a right-angled triangle, the square of the hypotenuse is sum of the squares of the two shorter sides. This can be written as $a^2 + b^2 = c^2$, where c is e hypotenuse.

1 Work out the lengths of the unknown sides.

(a) $h^2 = (4.7)^2 + (6.8)^2 = $ .... + ...

  $= $ .... cm^2

  $h = \sqrt{.............}$

  $h = $ .... cm

4.7 cm h

6.8 cm

(2 marks)

(b) $y^2 = 7.1^2 - 5.2^2 = $ .... + ...

  $= $ .... cm^2

  So, $y = \sqrt{.............}$

  $y = $ .... cm

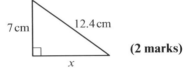

y 7.1 cm

5.2 cm

(2 marks)

(c) $x = $ .... cm

7 cm   12.4 cm

x

(2 marks)

2 A book of height 22 cm is standing upright in a bookcase. The top edge of another book with a height of 28 cm is resting against the top edge of the upright book. What is the distance between the bottom edges of the books?

> Draw and label a diagram which shows the information given as a right-angled triangle.

Distance = ... cm

(2 marks)

PROBLEM SOLVED!

3 The distance from Gloucester to Kidderminster is 44.1 miles due north. The distance from Gloucester to Luton is 111.4 miles due east. Calculate the direct distance between Kidderminster and Luton.

> You will need to use problem-solving skills throughout your exam – **be prepared!**

Kidderminster to Luton = ... miles

> Remember, north and east are at right-angles to each other.

(2 marks)

Guided

4 The council regulations for a park are that gaps between litter bins on a path must be less than 32 m. The map shows a park with paths at right-angles. Does it comply with the regulations?

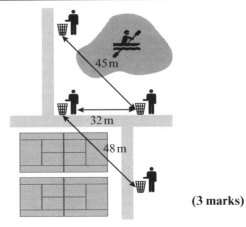

45 m

32 m

48 m

....

(3 marks)

Line segments

1   Work out the length of the line segment.

> You can find the length of a line segment between two points on a coordinate grid by using Pythagoras' theorem.

$x_B - x_A = 9 - 1 = 8$

$y_B - y_A = 8 - 3 = 5$

$AB^2 = 8^2 + \ldots\ldots\ldots^2 = \ldots\ldots\ldots$

$AB = \sqrt{\ldots\ldots\ldots}$

$AB = \ldots\ldots\ldots$ **(2 marks)**

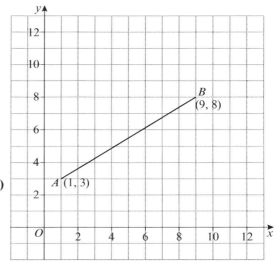

2   Calculate the length of the line segment.

$AB^2 = \ldots\ldots\ldots^2 + \ldots\ldots\ldots^2$

$ = \ldots\ldots\ldots$

$AB = \sqrt{\ldots\ldots\ldots}$

$ AB = \ldots\ldots\ldots$ **(2 marks)**

3   A line is drawn between coordinate D (2, 1) and coordinate E (7, −1). Calculate the length of line segment DE.

> Draw a right triangle using these coordinates, labelling the extra point at the right angle F.

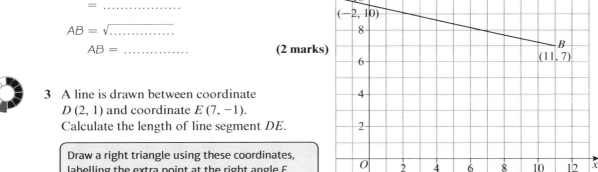

Line segment DE = ...

(2 marks)

4   A map is drawn at a scale of 1 square to 7 km. It shows three towns, Toptown (T), Crosstown (C) and Greentown (G).

(a)   Calculate the distance from Crosstown to Greentown.

.... km   **(1 mark)**

(b)   Calculate the distance from Toptown to Greentown.

.... km   **(1 mark)**

(c)   Calculate the distance from Toptown to Crosstown.

.... km   **(2 marks)**

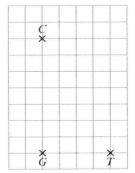

Trigonometry 1

> Trigonometric ratios are remembered using the acronym SohCahToa.
> $\sin\left(\frac{opp}{hyp}\right)$ $\cos\left(\frac{adj}{hyp}\right)$ $\tan\left(\frac{opp}{adj}\right)$

1 Work out the unknown side in the triangles, giving your answers to 3 s.f.

(a)

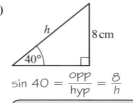

h 8 cm   40°

$$\sin 40 = \frac{opp}{hyp} = \frac{8}{h}$$

> Make h the subject of the equation.

$$h = \frac{8}{\sin 40}$$

$h = $ .... cm   **(3 marks)**

(b)

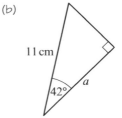

11 cm   42° a

> It is a good idea to label sides adj, opp and hyp.

.... $42° = \frac{adj}{hyp} = \frac{a}{11}$

$a = 11 \times$ ....

$a = $ .... cm   **(3 marks)**

2 Safety instructions on a ladder say that the base should make an angle no greater than 70° with the floor when used. If the ladder is 5 m long, what is the maximum height it can reach against a wall?

> You will need to use problem-solving skills throughout your exam – **be prepared!**

1. Draw a triangular diagram showing the ladder, wall and ground.
2. Label the three sides of the triangle a, b and c.
3. Mark the angle of 70° between the ground and the ladder.
4. Taking the ladder to be 5 m long, use **SohCahToa** to calculate the height of the ladder up the wall.

Maximum height = ... m   **(2 marks)**

3 The angle of elevation from the top of flagpole **A** to flagpole **B** is 27°. Flagpole **A** is 6 m tall and flagpole **B** is 6.9 m tall. How far apart are they to the nearest cm?

> You will need to use problem-solving skills throughout your exam – **be prepared!**

> Draw a triangle to show the angle of elevation from A to B. Label the information you know on the triangle.

Distance apart = ... cm   **(2 marks)**

4 The angle of elevation of the sun in London at a certain time one day was 50°. The shadow cast by a building on this day at this time was 31.24 m long. Calculate the height of the building.

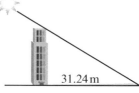

31.24 m

Height of building = ... m (4 s.f.)   **(3 marks)**

Trigonometry 2

5 Explain why the ratio of sides $a : b : c$ is equal to the ratio of sides $p : q : r$.

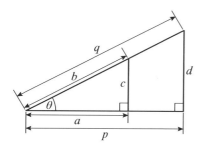

> Use what you know about similar triangles to show that the sine, cosine and tangent ratios are the same in the two triangles.

....

.... **(3 marks)**

6 Calculate the angles marked θ giving your answer to 3 s.f.

> To calculate an unknown angle, you must use the inverse function on your calculator.

(a)

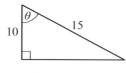

$\cos \theta = \left(\dfrac{\ldots\ldots}{\ldots\ldots} \right)$

Angle $\theta = \cos^{-1}\left(\dfrac{10}{\ldots\ldots} \right)$ (4 s.f.)

$\theta = $ ....

(3 marks)

(b)

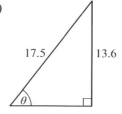

Angle $\theta = $ .... (4 s.f.)

(3 marks)

7 Two right-angled triangles share a side as shown in the diagram.

(a) Calculate angle θ.

Angle $\theta = $ .... (3 s.f.) **(2 marks)**

(b) Calculate the length of side q

(i) using trigonometry

Side $q = $ .... cm (3 s.f.) **(2 marks)**

(ii) using Pythagoras' theorem to check your answer to (i).

Side $q = $ .... cm (3 s.f.) **(2 marks)**

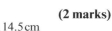

14.5 cm

8 cm

38.5°

8 An aircraft was 900 m from a point on the ground and this point on the ground was 5 km away from the runway of an airport.
Calculate the angle of elevation from the runway to the aircraft.

> Draw a diagram labelling the known sides and unknown angles.

Angle of elevation = ... (3 s.f.) **(3 marks)**

∪olving trigonometry problems

1 The diagrams show two triangles illustrating some special side and angle values.

(a) Complete the unknown sides and angles.

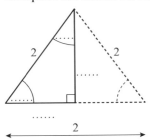

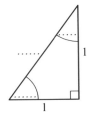

> Use what you know about equilateral and isosceles triangles. Give lengths in surd form for accuracy.

(6 marks)

(b) Complete the table using surds where necessary.

	0°	30°	45°	60°	90°
sin		$\frac{1}{2}$		$\frac{\sqrt{3}}{3}$	
cos			$\frac{\sqrt{2}}{2}$		
tan			1		Not defined

(5 marks)

2 Work out the length in cm of side y. Give your answer in surd form.

$$\cos 60° = \frac{y}{\sqrt{5}} \text{ and } \cos 60° = \frac{1}{2}$$

$$\frac{y}{\sqrt{5}} = \frac{1}{2}$$

$2y = \dots\dots\dots\dots\dots$

$\text{side } y = \dots\dots\dots\dots\dots \text{ cm}$

> Write cos 60° out in fraction form using information given in the diagram.

(3 marks)

3 Work out the size of angle θ in each triangle.

(a)

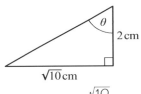

$\tan \dots\dots = \frac{\sqrt{10}}{\dots\dots}$

Angle $\theta = \dots\dots\dots\dots$ (3 s.f.)

(2 marks)

(b)

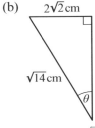

$\sin \theta = \frac{2\sqrt{2}}{\dots\dots}$

Angle $\theta = \dots\dots\dots\dots$ (3 s.f.)

(2 marks)

4 Two pieces of paper, one a square, the other a rectangle, are folded diagonally. The lengths of both diagonals are identical. Work out to 3 s.f.

(a) side x \dots\dots\dots\dots\dots\dots   **(2 marks)**

(b) side y \dots\dots\dots\dots\dots\dots   **(2 marks)**

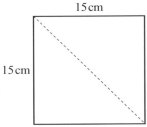

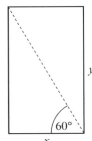

Measuring and drawing angles

1 Measure the size of the angles.

> Check whether the angle is acute or obtuse.

(a)

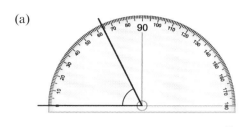

.... **(1 mark)**

(b)

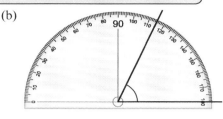

.... **(1 mark)**

(c) The angle shown is a reflex angle. What could its measurement be?

> There are two answers to this question. Drawing the possible angles will be a help.

(i) .... **(1 mark)**

(ii) .... **(1 mark)**

2 Add a line to form the following angles and label them.

> Place your protractor on the lines and mark off the angles.

(a) 42°

(b) 331°

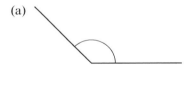

_____ **(1 mark)**

(1 mark)

3 Measure the following angles using a 180° protractor only.

(a)

.... **(1 mark)**

(b)

.... **(1 mark)**

4 Make an accurate drawing of the following pentagon.

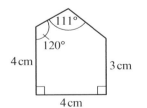

> You will need to use problem-solving skills throughout your exam – **be prepared!**

(4 marks)

95

Measuring lines

1 Measure each line in the shapes. Include units of measurement in your answer.

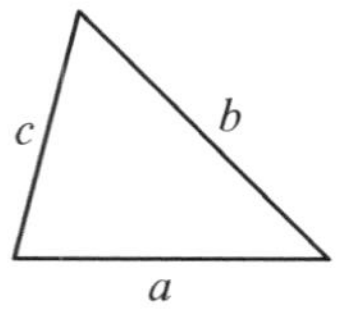

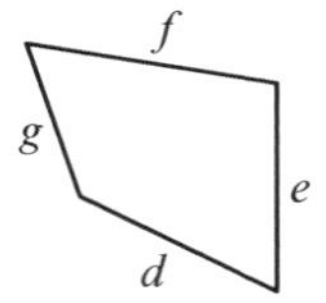

Place your ruler along the line you wish to measure, with 0 over the left corner of the line, and read off the value on the ruler scale.

(a) .... **(1 mark)**   (b) .... **(1 mark)**

(c) .... **(1 mark)**   (d) .... **(1 mark)**

(e) .... **(1 mark)**   (f) .... **(1 mark)**

(g) .... **(1 mark)**

2 Draw an accurate triangle with the following measurements, labelling the sides a, b and c.

$a = 4.2$ cm, $b = 5.7$ cm, $c = 35$ mm.

(3 marks)

3 Mark a point P that is exactly $\frac{3}{4}$ the distance from A to B

A ————————————————————— B

Measure the length of the line and multiply it by $\frac{3}{4}$.

(2 marks)

4 If the height of the door is approximately 2.3 m, estimate the height of the building.

.... m   **(2 marks)**

5 The great pyramid at Giza is approximately 147 m tall. A car (not drawn to scale) is roughly 86 times smaller.

Estimate the height of the car.

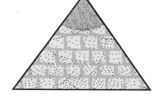

.... m   **(2 marks)**

Plans and elevations

1 If the diagram is a net of a polyhedron, name the polyhedron. If not, mark as false.

In (a) remember the formula for the circumference of a circle is $C = \pi D$

(a)

.... **(1 mark)**

(b)

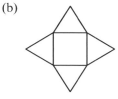

.... **(1 mark)**

(c)

.... **(2 marks)**

2 The diagram shows 13 small cubes arranged into an irregular 3D object.

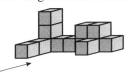

(a) Complete this drawing of the front elevation of the object.

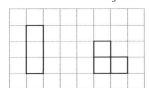

(2 marks)

(b) Complete this drawing of the plan of the object.

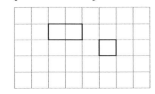

(2 marks)

3 The plan, front and side elevations of a 3D object are given. Complete the isometric drawing of the 3D object.

Plan   Front   Side

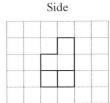

(2 marks)

4 The plan and front elevation of a 3D object are given. Draw the side elevation of this object.

Plan   Front   Side

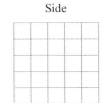

(2 marks)

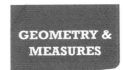
Had a go ☐   Nearly there ☐   Nailed it! ☐

Scale drawings and maps

1 The following lengths on maps are measured with a rule. Use the scales to calculate their actual lengths.

(a) 4.5 cm; Scale: 1.5 cm = 1 m

Actual length = 4.5 ÷ .... × ....

.... m   **(2 marks)**

(b) 54 mm; Scale: 1 cm = 1 km

Length = 54 mm = 5.4 cm

Actual length = ... km   **(2 marks)**

(c) 7.6 cm; Scale: 85 mm = 2 km

2 × .... = ... (2 d.p.)

> Find the multiple that turns 85 mm into 760 mm. 760 ÷ 85 = 8.94 (2 d.p.).

(2 marks)

2 What actual distance in km is represented by the following distances measured on a map?

> When converting, you need to decide whether to do a multiplication or a division.

(a) Distance on map = 18 cm; scale = 1 : 90 000

Actual distance = ... cm × 90 000

= .... cm

.... cm ÷ 100 = ... m

.... m ÷ 1000 = ... km

.... km   **(2 marks)**

(b) Distance on map = 9.7 cm; scale = 1 : 100 000

Actual distance = ... cm × 100 000 = ... cm

.... cm ÷ 100 = ... m

.... m ÷ 1000 = ... km

.... km   **(2 marks)**

(c) Distance on map = 4 cm; scale = 1 : 1 000 000

....   **(2 marks)**

3 Hamish is making a model of the aircraft carrier *Queen Elizabeth II*. The ship is 280 m long and he has decided to make a 1 : 72 scale model. Hamish has a modelling room which is 6 m long but says his room could be $\frac{2}{3}$ this size and the model could still be built in it. Explain why you either agree or disagree with Hamish's claim.

....

....   **(2 marks)**

4 The time taken to travel between Ouistreham and Portsmouth on a ferry is 5 hours 30 minutes. The ferry travels at a mean speed of 33 km/h. On board is an animated scale map of the journey, representing the distance as 2 m, and plotting the position of the ferry as it travels. If the map is on a scale 1 : 90.75, what is the speed of the animated ferry on the map? Give your answer in m/s to 1 s.f.

....   **(4 marks)**

Constructions 1

 1 Using just a pencil, ruler and compasses, construct a perpendicular bisector of the line AB.

> Open compasses to about $\frac{2}{3}$ the length of AB. With the point of the compasses on A, draw an arc passing through the line. Repeat this from point B. Draw a straight line through the points where the arcs intersect.

(2 marks)

 2 Use a ruler, pencil and compasses to construct a perpendicular to the line segment PR from point Q.

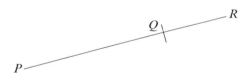

> Open your compasses, place the point on Q, and draw arcs on the line equidistant from Q. Use these two points, where the arcs cross the line, to produce the perpendicular line that cuts PR and passes through Q.

(2 marks)

 3 Using compasses, pencil and ruler, construct a perpendicular from the line segment XY through the point W.

Guided

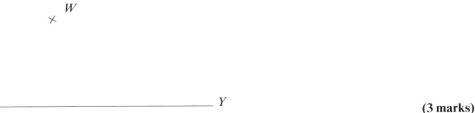

(3 marks)

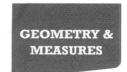
Constructions 2

4 Using just a pencil, ruler and compasses, construct a rhombus with side lengths 5 cm.

> 1. Draw a line 5 cm long.
> 2. Label the ends A and B.
> 3. Open your compasses to 5 cm and draw arcs from both A and B.
> 4. The two arcs will meet at two points, one on either side of the line AB.
> 5. Join all four points.

(2 marks)

5 Draw a line AC that meets at angle A on the line AB at 45°. You can use a protractor to do this. Using compasses and a ruler only, bisect angle A. Do not use a protractor for this.

> Do not rub out the arcs you make when using your compasses.

A ——————————— B

(2 marks)

6 Using compasses, pencil and ruler, construct a scalene triangle with angles 90°, 60° and 30°.

> Start by drawing a line of say 4 cm and draw in a perpendicular from the right hand side. If the perpendicular is larger than the first line it will be more helpful.

(2 marks)

7 Draw three adjoining lines AB, BC and CD of a regular octagon. Use only a ruler, pencil and compasses.

(3 marks)

Loci

1 Draw the locus of all points that are exactly 2 cm from the line *AB*.

A ——————— B

> Draw circles of radius 2 cm with centres *A* and *B* and join them with horizontal lines.

(2 marks)

2 A guard dog is chained to the corner of a rectangular 8 m × 4 m building at *B*. The chain is 5 m long. Shade the area around the building that the dog can reach. The scale of the drawing is 1 cm = 2 m.

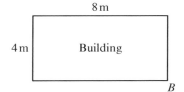

> 1. Open your compasses to the equivalent of 5 m.
> 2. Draw an arc with centre *B* spanning the area outside the rectangle.
> 3. Look back at the question. Notice where the building is. Which area can the dog reach?

(3 marks)

3 Shade the area inside the triangle *ABC* that is closer to line *AC* than *AB* but is less than 3 cm from point *B*.

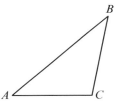

> 1 Bisect angle *BAC*.
> 2 Draw an arc of radius 3 cm centred on point *B*.
> 3 Shade the area that is beneath the bisector line you drew, and also inside the arc centred on *B*.

(4 marks)

Guided

PROBLEM SOLVED!

4 Three major supermarkets are shown on a map as *L*, *M* and *W*. Leila lives within 10 km of all three. Shade in the area on the map where her home might be.

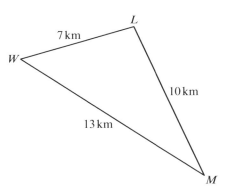

> You will need to use problem-solving skills throughout your exam – **be prepared!**

(2 marks)

Had a go ☐ **Nearly there** ☐ **Nailed it!** ☐

Bearings

1 Write the bearings of

(a)

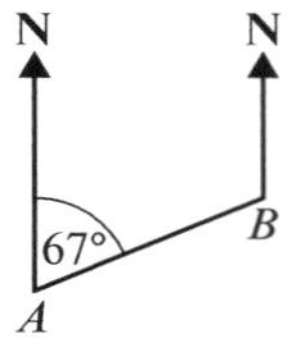

> You must always face north from the starting position. Bearings are always written with three digits, even before decimal points.

(i) *B* from *A*

67° = ...°

> Use the diagram and rewrite the angle as a bearing.

(ii) *A* from *B*

180° + ...° = ...° **(2 marks)**

(b)

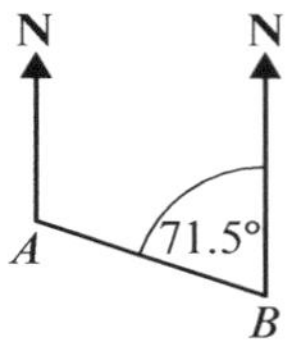

> Start facing north at *B*. Do you measure the angle clockwise or anticlockwise?

(i) *B* from *A*

= ....°

(ii) *A* from *B*

= ....° **(2 marks)**

2 Draw a line on a bearing of

(a) 036°

N

> Place your protractor with 0° aligned with North and draw a line 36° in a clockwise direction.

(1 mark)

(b) 185°

N

> Remember 185° clockwise from North is the same as 175° anti-clockwise from North.

(1 mark)

3 Three towns, *A*, *B* and *C* are shown on a map which is drawn on the scale 1 cm = 3 km.

> You will need to use problem-solving skills throughout your exam – **be prepared!**

(a) Measure the bearing of *C* from *B*.

....

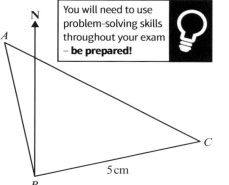

(1 mark)

(b) A fourth town, *D*, is on a bearing of 192° from town *C* and a bearing of 097° from town *B*. Plot town *D*.

(c) What is the approximate distance to the nearest km of town *D* from town *A*?

....

(2 marks)

(1 mark)

Circles

1 Label the four parts of a circle on the diagram.

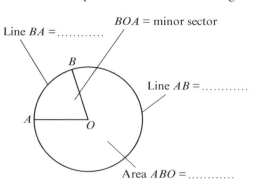

Line BA = ...

BOA = minor sector

Line AB = ...

Area ABO = ...

> If a minor sector is identified in a circle, there is always another sector that goes with it to make up the whole circle, called the major sector. If both sectors are the same size they are semicircles.

(3 marks)

2 Work out the circumference of the circles, giving your answers in terms of π.

(a)

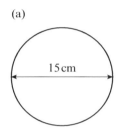

15 cm

(b)

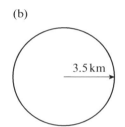

3.5 km

> Circumference (C) = $2\pi r$ or πD

(a) $D = 15$

$C = \pi D = 15\pi$

$C = ... $ cm   **(2 marks)**

(b) $C = 2\pi r = 2 \times r \times \pi = 2 \times .... \times \pi$

$C = ... \pi$ **(2 marks)**

3 Work out the radii of the circles with the following circumferences (C). Give exact values for parts (a) and (b).

> $D = 2r$

(a) $C = 27$ cm

$C = \pi D = 2\pi r = 27$

So $r = \dfrac{27}{......} = \dfrac{......}{......}$ cm   **(2 marks)**

(b) $C = 90$ m

$C = \pi D = 2\pi r = 90$

So $r = \dfrac{.........}{2\pi} = \dfrac{......}{......}$ m   **(2 marks)**

 Guided

4 Work out the perimeters (P) of the following shapes in terms of π.

(a)

14 cm

(b)

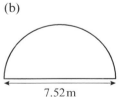

7.52 m

(a) ....   **(3 marks)**   (b) ....   **(3 marks)**

 Guided

5 A carousel on a fairground ride has a diameter of 9 m from one horse to another opposite. How many cycles would it take for a horse to have travelled 2 km? (3 s.f.)

....   **(3 marks)**

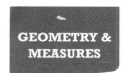
Area of a circle

1 Work out the area of the following circles in terms of π.

(a)

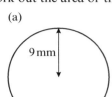

9 mm

(b)

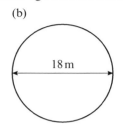

18 m

> The area of a circle is found using the formula $A = \pi r^2$

(a) $A = \pi r^2$

$A = \pi(\ldots\ldots\ldots^2) = \ldots\ldots\ldots \pi \ mm^2$

(b) $A = \pi(\ldots\ldots\ldots^2) = \ldots\ldots\ldots \pi \ m^2$

> From the diagram, $r = 9$.

> $A = \pi r^2$. Be careful to use the correct value for r.

(2 marks)

(2 marks)

2 Work out the areas of the circle parts.
Give an exact answer in terms of π.

> Area of a quadrant is 4 times smaller than the area of a circle.

(a) $A = \pi r^2$

Area of circle $= 400\pi$

Area of quadrant $= 400\pi \div 4$

Area of quadrant $= \ldots\ldots\ldots\ldots$

(b) $A = \dfrac{\pi \times 8^2}{\ldots\ldots\ldots\ldots}$

$A = \ldots\ldots\ldots\ldots\ldots$

(c) $\ldots\ldots\ldots\ldots\ldots$

(a)   (b)

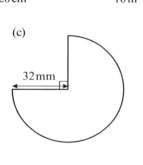

20 cm   16 m

$r = 20$

(c)

32 mm

(2 marks)

(2 marks)

(2 marks)

> Guided

3 Work out the shaded areas. Give answers to 3 s.f.

> Area of annulus = larger circle area − smaller circle area

(a)

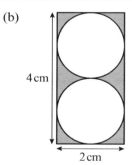

20 mm

10 mm

(b)

4 cm

2 cm

$\ldots\ldots\ldots\ldots\ldots\ldots$ **(3 marks)** $\ldots\ldots\ldots\ldots\ldots\ldots$ **(3 marks)**

Sectors of circles

1 Work out the arc lengths of the following sectors of circles. Give an exact answer in terms of π.

(a)

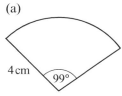

4 cm 99°

(b)

162° 5 cm

> The circumference of a circle is found using the formula $C = 2\pi r$.
> The angle of the sector is a fraction of 360°.
> Multiply the circumference by this fraction to find the arc length.

(a) $r = 4$ cm

$C = 2 \times 4 \times \pi = 8\pi$

Arc length $= \dfrac{\ldots}{360} \times 8\pi = \dfrac{\ldots \times 8\pi}{360}$

Arc length $= \ldots\ldots\ldots\ldots\ldots$ cm

> You can cancel the fraction before calculating your answer.

(2 marks)

(b) Arc length $= \dfrac{\ldots\,\pi \times 162}{360}$

Arc length $= \ldots\ldots\ldots\ldots\ldots$ cm

(2 marks)

2 Work out the perimeters of the circle sectors. Give your answer to 3 s.f.

Guided

(a)

$\ldots\ldots\ldots\ldots\ldots\ldots\ldots\ldots$ **(3 marks)**

> Perimeter of sector = arc length + two side lengths.

(a)

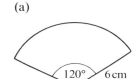

120° 6 cm

(b)

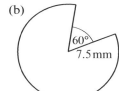

60°
7.5 mm

(b)

$\ldots\ldots\ldots\ldots\ldots\ldots\ldots\ldots$ **(3 marks)**

3 Work out the areas of the following sectors of circles.

(a)

51° 4.5 mm

(b)

135°
23 cm

> Area of a circle: $A = \pi r^2$
> Area of a sector with radius r and angle $\theta°$:
> $A = \dfrac{\theta}{360}\pi r^2$

(a) $\dfrac{51}{360} \times \pi \times \ldots\ldots \times \ldots\ldots\ldots = A = \dfrac{\ldots\ldots}{360}\pi \times \ldots\ldots\ldots \times \ldots\ldots\ldots$

$= \dfrac{\ldots\ldots \times \ldots\ldots \times \ldots\ldots \times \pi}{360}$

$A = \ldots\ldots\ldots\ldots\ldots$ mm²

(3 marks)

(b) $A = \dfrac{\ldots\ldots\ldots \times (\ldots\ldots\ldots)^2 \times \pi}{360} = \dfrac{\ldots\ldots\ldots}{\ldots\ldots\ldots}$

Area $= \ldots\ldots\ldots\ldots\ldots$ cm²

> Make sure you use the right fraction.

(3 marks)

Cylinders

1 Work out the volumes of the following cylinders to 3 s.f.

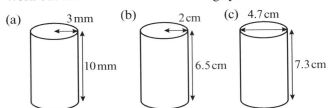

(a)   3 mm   (b)   2 cm   (c)   4.7 cm

     10 mm   6.5 cm   7.3 cm

(a) $V = \pi r^2 h = \pi \times 3^2 \times 10 = $ ....

....

> The volume of a cylinder is found using the formula $V = \pi r^2 h$.
> h = vertical height of the cylinder.

(2 marks)

(b) $V = \pi r^2 h = \pi \times 2^2 \times 6.5 = $ ....

....

(2 marks)

(c) $V = \pi r^2 h = \pi \times $ .... $^2 \times $ .... $ = $ ....

....

(2 marks)

2 Work out the surface areas of the cylinders. Give your answer in terms of π.

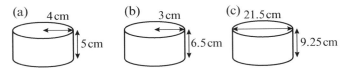

(a)   4 cm   (b)   3 cm   (c) 21.5 cm

     5 cm   6.5 cm   9.25 cm

> The surface area of a cylinder is calculated from the formula
> $SA = 2\pi r^2 + 2\pi rh$
> $SA = 2\pi r(r + h)$, where h = vertical height.

(a) $SA = 2\pi r^2 + 2\pi rh = 2\pi r(r + h)$

    $SA = 2 \times \pi \times $ .... $\times ($ .... $ + $ ....$)$

    $SA = $ .... π cm^2

(3 marks)

(b) $SA = 2\pi r(r+h) = 2 \times \pi \times 3 \times (3 + 6.5)$

    $SA = $ ....

(3 marks)

(c) ....

(3 marks)

3 The diagram shows a cuboid and a cylinder. The cylinder is exactly 3 times the volume of the cuboid.

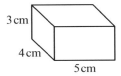 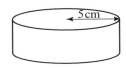

3 cm   5 cm

4 cm

     5 cm

> Work out the volume of the cuboid, times this by 3 and then set this equal to the cylinder volume formula.

(a) Calculate the height of the cylinder to 3 s.f.

    ....

(3 marks)

(b) Which of the two objects has the greater surface area?

    ....

(3 marks)

Volumes of 3D shapes

1 Work out the volumes of the following 3D figures. Give your answers to 3 s.f.

(a)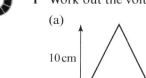
10 cm
10 cm

(b)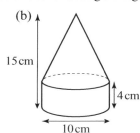
15 cm
4 cm
10 cm

(c)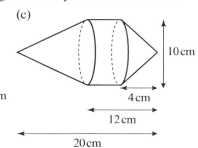
10 cm
4 cm
12 cm
20 cm

(a) $V = \dfrac{\pi r^2 h}{3}$

$V = \dfrac{\pi \times 5 \times \text{......} \times \text{......}}{3}$

$V = \text{....................} \ cm^3$

Volume of a cone: $V = \dfrac{\pi r^2 h}{3}$
h = vertical height of the cylinder.

(2 marks)

(b) ....

Calculate the volumes of the cone and the cylinder separately and then add them together.

(2 marks)

(c) ....

(3 marks)

2 Object (a) is a sphere. Object (b) is a hemisphere on a cylinder. Object (c) is a hemisphere on a cone. Work out the volume of each, giving your answer in terms of π.

(a)
16 cm

(b)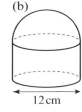
12 cm
12 cm

(c)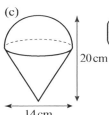
20 cm
14 cm

Volume of a sphere: $V = \dfrac{4}{3}\pi r^3$

(a) $V = \dfrac{4}{3}\pi r^3$

$V = \dfrac{4}{3}\pi \times \text{.........} \times \text{.........} \times \text{.........}$

$V = \text{.........} \ cm^3$

(3 marks)

(b) Calculate the volumes of the hemisphere and the cylinder separately and add them together. They both have the same radius (r) value, so work that out first.

....

(3 marks)

(c) ....

(3 marks)

3 A cylinder and a cone are shown in the diagram and both have the same volume. Calculate the height of the cone, y.

Guided

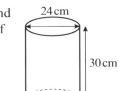
24 cm
30 cm

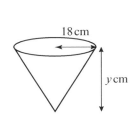
18 cm
y cm

....

(4 marks)

107

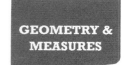
Had a go ☐   **Nearly there** ☐   **Nailed it!** ☐

Surface area

1 The surface area of a square cross-sectioned cuboid is $192\,cm^2$. Find the value of n.

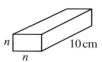

n $10\,cm$ n

> A cuboid's surface area is the sum of the areas of the three opposing vertical faces.

.... **(2 marks)**

Guided

2 A cuboid is labelled as in the diagram.

Calculate side n.

$120\,cm^2$

$6\,cm$ $72\,cm^2$

n

.... **(2 marks)**

3 Calculate the surface area of the cylinder in the diagram. Give your answer in terms of π.

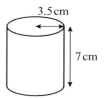

$3.5\,cm$

$7\,cm$

> The formula for the surface area of a cylinder (where r = radius of the circular face and h = the height of the cylinder) is: $2\pi r^2 + 2\pi rh = 2\pi r(r + h)$

.... **(3 marks)**

4 A cylinder has a radius of $7.5\,cm$ and a surface area of $255\pi\,cm^2$. Calculate its height.

.... **(3 marks)**

> You will need to rearrange the formula to solve this.

5 Calculate the surface area of the cones with the following heights and radii.

(a) $r = 10\,cm$, $h = 10\,cm$

....

(b) $r = 1.3\,m$, $h = 5.2\,m$

h

r

> The surface area of a cone is: $(\pi \times r^2) + (\pi \times r \times s)$ where s is the slant height given by $\sqrt{h^2 + r^2}$

(2 marks)

.... **(2 marks)**

6 The surface area of a sphere in terms of π is given as $4\pi r^2$ where r = radius of the sphere.

Guided

(a) From this formula of the surface area of a sphere, work out the formula for the surface area of a hemisphere.

.... **(4 marks)**

(b) Use your formula to find the surface area of a hemisphere with a radius of $12.3\,cm$.

.... **(3 marks)**

Similarity and congruence

1 The following shapes are not drawn to scale. Write down which shapes are

> Use Pythagoras' theorem.

(a) similar.   (b) congruent.

.... **(2 marks)**   .... **(2 marks)**

2 (a) On the grid, draw a shape congruent to the one shown.

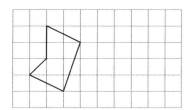

> Two shapes are congruent if they are identical in both shape and size, even if one of them has been reflected or rotated.

(b) Draw a triangle that is similar, but not congruent, to the one shown below. You may use a ruler and compasses but not a protractor. Show your working.

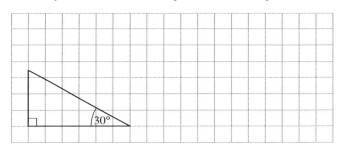

> Work out the ratio between the shortest two sides.

(4 marks)

3 Greg, Bethan and Rick are discussing the three rectangles below, which are not drawn to scale. Bethan says rectangles **A** and **C** are congruent and are the only congruent ones. Greg disagrees and says that none of them are congruent. Rick thinks all three rectangles are congruent to each other. Write down who is correct, giving your reasons.

....

.... **(3 marks)**

Had a go ☐ **Nearly there** ☐ **Nailed it!** ☐

Similar shapes

1 Shapes **A** and **B** are similar and have one line of symmetry as shown in **A**.

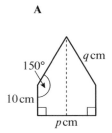

A

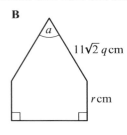

B

> Similar shapes have identical angles. Use this and what you know about interior angles of polygons to help you answer this question.

(a) Calculate angle a.

.... **(2 marks)**

(b) Calculate side p.

.... **(2 marks)**

(c) Calculate side q.

.... **(2 marks)**

(d) Calculate side r.

.... **(2 marks)**

2 Triangles **A** and **B** are mathematically similar.

> Use $0.6 \times \theta = 108$ to work out θ.

(a) Calculate angle a.

.... **(2 marks)**

(b) Calculate angle b.

.... **(2 marks)**

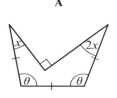

A

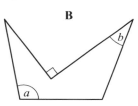

B

(c) Calculate side x.

.... **(2 marks)**

Guided

3 Shape **A** started out as a regular pentagon, then the top vertex was repositioned so as to form a right angle as shown. Shape **B** is mathematically similar to shape **A**.

(a) Calculate angle a.

.... **(3 marks)**

(b) Calculate angle b.

.... **(3 marks)**

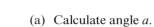

A   B

4 The diagram shows an irregular hexagon with a line of symmetry AB. Draw a similar shape to this using the line CD as the line of symmetry.

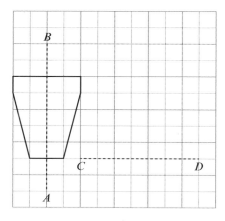

(2 marks)

Congruent triangles

1 Prove that the two triangles shown are congruent.

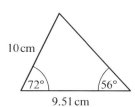 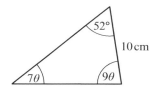

> Show that the two triangles have equal angles and equal side lengths.

(3 marks)

2 Triangles **A**, **B** and **C** are congruent. Calculate the angles and label them on triangle **C**.

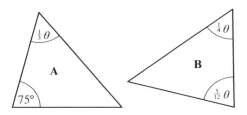

> Label angles on both **A** and **B** then compare. **B** will be the easier one to label first.

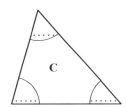

(3 marks)

3 Is it possible that quadrilaterals **A** and **B** below are not congruent? Explain your answer.

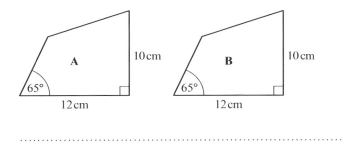

....

.... **(3 marks)**

Vectors

1 Draw, label and join the following vectors.

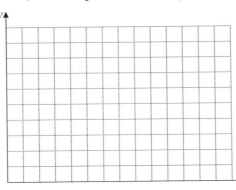

The starting point of any vector can be any point in space.

(a) $\overrightarrow{AB} = \begin{pmatrix} 2 \\ 3 \end{pmatrix}$ **(1 mark)**

(b) $\overrightarrow{BC} = \begin{pmatrix} -5 \\ -3 \end{pmatrix}$ **(1 mark)**

2 Write each vector as a column vector.

(a) $\overrightarrow{AB}$ (d) $\overrightarrow{AD}$

$\begin{pmatrix} .... \\ 2 \end{pmatrix}$ $\begin{pmatrix} ...... \\ ...... \end{pmatrix}$

(b) $\overrightarrow{BC}$ (e) $\overrightarrow{AC}$

$\begin{pmatrix} 2 \\ ...... \end{pmatrix}$ $\begin{pmatrix} ...... \\ ...... \end{pmatrix}$

(c) $\overrightarrow{DC}$ (f) $\overrightarrow{BD}$

$\begin{pmatrix} ...... \\ ...... \end{pmatrix}$ $\begin{pmatrix} ...... \\ ...... \end{pmatrix}$

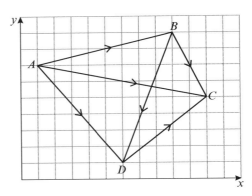

(6 marks)

3 The diagram shows a parallelogram *PQRS*.
Write down the following vectors in terms of **q** and **r**.

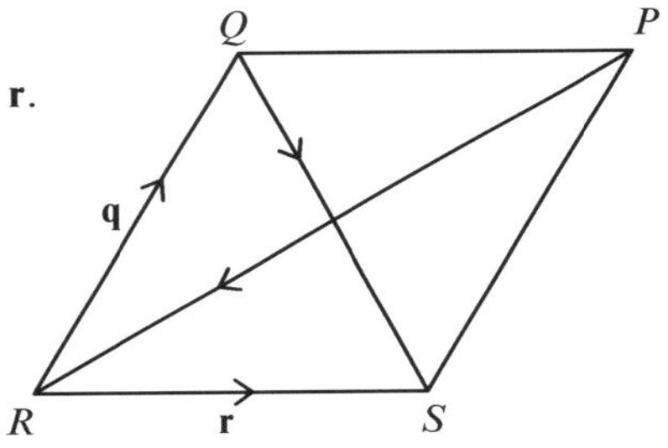

(a) $\overrightarrow{RP}$

q + ...

(b) $\overrightarrow{PR}$

....

(c) $\overrightarrow{SQ}$

q − ....

(d) $\overrightarrow{QS}$

.... **(4 marks)**

4 $\overrightarrow{AO} = \mathbf{a} + \mathbf{b}$, $\overrightarrow{OB} = 2\mathbf{a} - \mathbf{b}$

Write down the following vectors in terms of **a** and **b**.

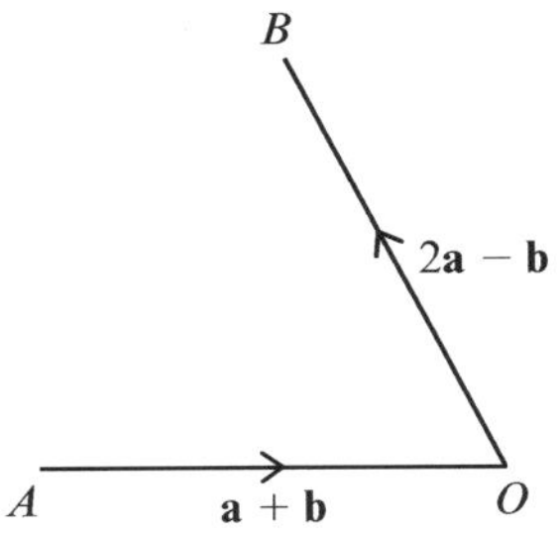

(a) $\overrightarrow{AB}$

....

(b) $\overrightarrow{BA}$

....

(2 marks)

(2 marks)

Problem-solving practice 1

1 Calculate

   (a) angle x

     ....

   (b) angle y

     ....

   (c) angle z

     ....

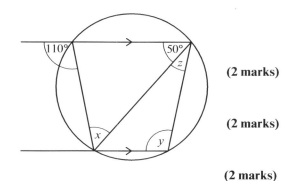

(2 marks)

(2 marks)

(2 marks)

2 Show that cylinder **A** can be filled with water by exactly 2 fills from cylinder **B**.

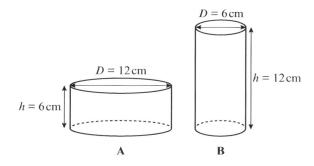

(3 marks)

3 The diagram shows a rectangle **A**. It is rotated 90° clockwise about (0, 0) to the position **B**. The rectangle **B** is then reflected in the line $y = -x$ and this rectangle is labelled **C**. Describe the single transformation that takes **A** to **C**.

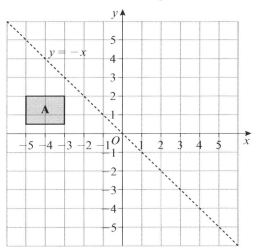

.... **(3 marks)**

113

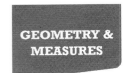
Had a go ☐   **Nearly there** ☐   **Nailed it!** ☐

Problem-solving practice 2

Guided

4 A regular pentagon *ABCDE* and two isosceles triangles *EDF* and *CDG* are shown. Angle *FDG* = *x*°. Calculate angle *FDG*.

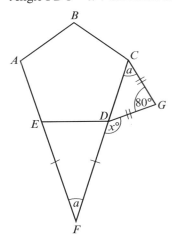

....

(4 marks)

5 A ship sails from port *A* on a bearing of 060° to port *B*. From there it takes a bearing of 110° to port *C* which is 112 km due east from port *A*. Calculate the distance to 3 s.f. between ports *B* and *C*.

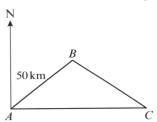

....

(4 marks)

6 A 30 cm long conical peg with a profile angle of 20° exactly fits a symmetrical conical hole in a piece of wood. The diameter of the larger entry hole is 2 cm and that of the smaller exit hole is 1 cm. Calculate the thickness of the piece of wood (*x*).

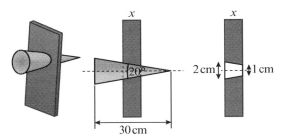

....

(5 marks)

114

Two-way tables

1 65 sports teams travelled to tournaments in Germany.
The two-way table shows some information about the visits.

	Berlin	Bielefeld	Bremen	Total
Football	....	11	....	31
Tennis	....	....	9	....
Total	30	14	....	65

> Look for a row or column with two results first. Use the total figures to work out the missing numbers

Complete the two-way table.   **(3 marks)**

2 84 children had an activity afternoon at school at the end of term. The two-way table shows their activity choices.

> **Guided**

	Printing	Painting	Pottery	Total
Girls	20	17	....	44
Boys	....	16	....	....
Total	....	....	9	84

Complete the two-way table.   **(3 marks)**

3 The two-way table shows the holiday destinations of different age ranges for 150 randomly chosen people.

	Europe	UK	Neither	Total
Under 30	29	....	3	37
30–55	31	....	19	....
Over 55	....	....	....	49
Total	88	19	....	150

(a) Complete the two-way table.

                                                               (3 marks)

(b) How many people who were at least 30 years old travelled to Europe?

> Look in the Europe column. Add up all the people who were **not** under 30.

....   **(1 mark)**

(c) Gregor says that the group in the 'Neither' column did not have a holiday.
   Explain why he is wrong to assume this.

> Are Europe and the UK the only possible holiday destinations?

....   **(1 mark)**

Had a go ☐ Nearly there ☐ Nailed it! ☐

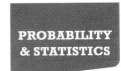

Pictograms

1 The pictogram shows the number of houses built by a construction company in four months of one year.

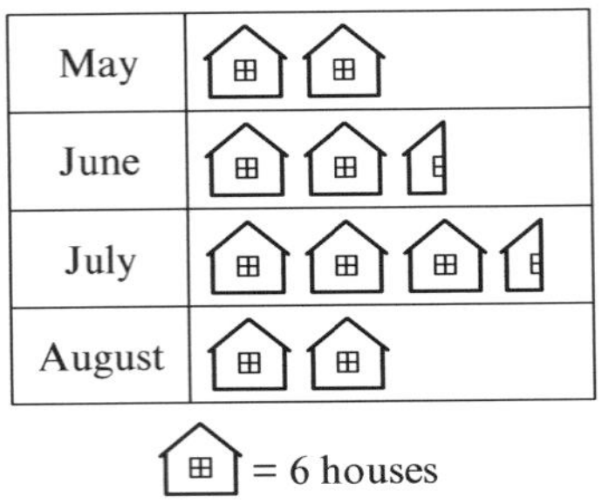

(a) The difference between the number of houses built in June and July is equal to half the number of houses built in May. Is this statement true or false?

Total built in July = ... × 3.5 = ...

Total built in June = 6 × .... = ...

Difference = ... − ....

Half the number built in May = $\frac{\text{......} \times 2}{\text{......}}$

The statement is .... because:

.... **(2 marks)**

(b) In the first four months of the year the company built $\frac{3}{5}$ as many houses as they did during the time shown in the pictogram. Calculate how many houses they built from January to April.

Number in May = ...

Number in June = ...

Number in July = ...

Work out the total number of houses in the pictogram and then find $\frac{3}{5}$ of this number.

Use your answers from (a) above. (Remember that you halved the amount for May!)

Number in August = 6 × .... = ...

Total in May to August = ... + ... + ... + ... = ...

Total in January to April = $\frac{\text{.........}}{\text{.........}}$ × .... = ... **(2 marks)**

(c) In the last four months of the year, the construction company built $\frac{9}{16}$ as many houses as they did in the first eight months. How many houses were built in the whole year?

Calculate the number of houses built in the first eight months.

Totals from part (b): .... +......... = ...

$\frac{9}{16}$ × .... = ...

.... + ... = ...

PROBLEM SOLVED!

You will need to use problem-solving skills throughout your exam – **be prepared!**

(3 marks)

Guided

2 The pictogram shows the number of visitors to a service station on the M40 between 05:00 and 06:00 on 5 days in one week.

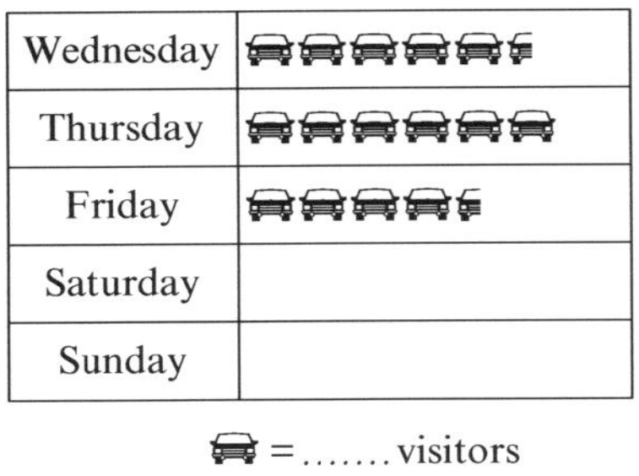

(a) The total number of visitors on Thursday and Friday was 420. Using this information, complete the key. **(2 marks)**

(b) How many visitors were there on Wednesday?

.... **(2 marks)**

(c) There were 100 visitors on Saturday, and 60 on Sunday. Use this information to complete the pictogram. **(2 marks)**

Bar charts

1 There are four houses in a school, and all houses have about the same number of students. Students were selected at random for a survey. The bar chart shows information about the selections from Hood and Rodney houses.

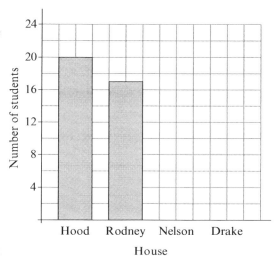

(a) A further 19 of the students selected were from Nelson house, and 18 of them were from Drake. Use this information to complete the bar chart.

(1 mark)

(b) True or false: 'The bars are not equal, so the students cannot have been chosen at random'? Explain your answer.

> Think of each bar as a fraction of the total. How similar are the values?

....   **(2 marks)**

2 Five students recently took tests in English and German. The bar chart shows some of the results.

(a) The sum of all marks for English was 180 and the sum for German was 170. If Ineke scored $\frac{1}{6}$ of all the marks for English and $\frac{1}{3}$ of the remaining marks for German,

(i) complete the bar chart for Ineke

(ii) complete the bar chart for Wim.

> Work out $\frac{1}{6}$ of 180 for the English mark. Calculate the remaining German marks and divide by 3.

(2 marks)

(2 marks)

(b) Who scored the best mark for German?

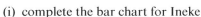

> Use the total number of marks to work out the number of marks for Wim.

(1 mark)

(c) Who scored the lowest result for English?

....

> You will need to use problem-solving skills throughout your exam – **be prepared!**

(1 mark)

Pie charts

1 A class of 40 pupils at a school are interviewed about how they travel to school each day. The results are shown in the table. Plot a pie chart to show this information.

Method of transport	Frequency
Walk	15
Bicycle	13
Bus	8
Car	4

$$\frac{360}{\dots} = \dots °$$

$15 \times \dots = \dots °$

$13 \times \dots = \dots °$

$8 \times \dots = \dots °$

$4 \times \dots = \dots °$

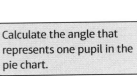
Calculate the angle that represents one pupil in the pie chart.

(3 marks)

2 Ieuan asked his friends to choose their favourite lesson from a choice of four. The results are shown in the table. Draw a pie chart of Ieuan's results.

Subject	Frequency
French	45
Science	19
PE	8
History	18

$$\frac{45}{\dots} \times 360° = \dots °$$

$$\frac{19}{\dots} \times 360° = \dots °$$

$$\frac{8}{\dots} \times 360° = \dots °$$

$$\frac{18}{\dots} \times 360° = \dots °$$

The angle representing each subject can be found by making a fraction with the frequency of that subject as the nominator and the total frequency as the denominator, then multiplying by 360°.

(3 marks)

Guided

3 A cricket academy has an intake of players from four counties, shown in the bar chart. The chart has been marked with some angles. Use the information given to calculate the intake from Sussex, Somerset and Surrey and complete the table.

County	Frequency
Lancashire	180
Sussex	
Somerset	
Surrey	

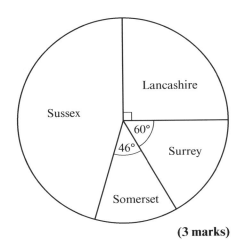

(3 marks)

Scatter graphs

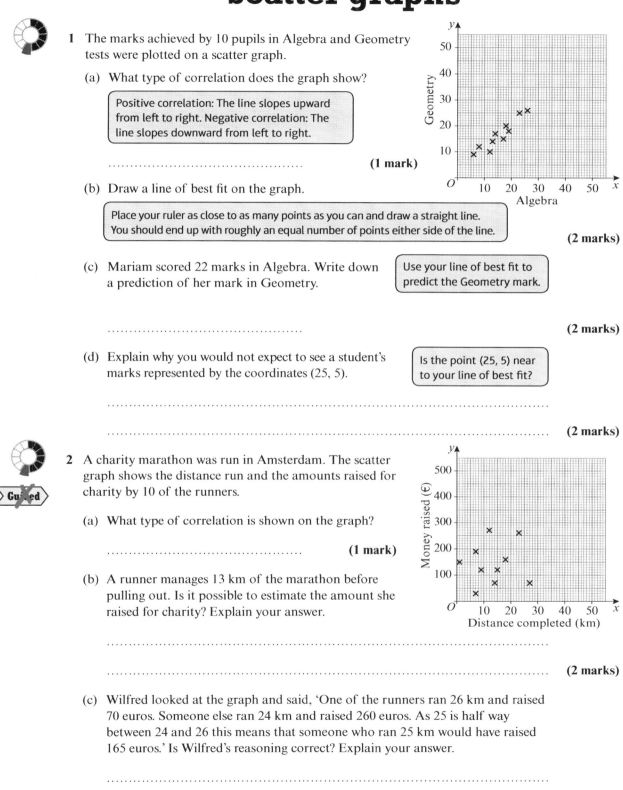

1 The marks achieved by 10 pupils in Algebra and Geometry tests were plotted on a scatter graph.

(a) What type of correlation does the graph show?

> Positive correlation: The line slopes upward from left to right. Negative correlation: The line slopes downward from left to right.

.... **(1 mark)**

(b) Draw a line of best fit on the graph.

> Place your ruler as close to as many points as you can and draw a straight line. You should end up with roughly an equal number of points either side of the line.

(2 marks)

(c) Mariam scored 22 marks in Algebra. Write down a prediction of her mark in Geometry.

> Use your line of best fit to predict the Geometry mark.

.... **(2 marks)**

(d) Explain why you would not expect to see a student's marks represented by the coordinates (25, 5).

> Is the point (25, 5) near to your line of best fit?

....

.... **(2 marks)**

2 A charity marathon was run in Amsterdam. The scatter graph shows the distance run and the amounts raised for charity by 10 of the runners.

(a) What type of correlation is shown on the graph?

.... **(1 mark)**

(b) A runner manages 13 km of the marathon before pulling out. Is it possible to estimate the amount she raised for charity? Explain your answer.

....

.... **(2 marks)**

(c) Wilfred looked at the graph and said, 'One of the runners ran 26 km and raised 70 euros. Someone else ran 24 km and raised 260 euros. As 25 is half way between 24 and 26 this means that someone who ran 25 km would have raised 165 euros.' Is Wilfred's reasoning correct? Explain your answer.

....

.... **(3 marks)**

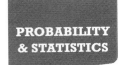

Had a go ☐ Nearly there ☐ Nailed it! ☐

Averages and ranges

1 The marks scored by 10 pupils in a test were as follows.
5, 9, 13, 8, 9, 11, 4, 15, 7, 9

Calculate

(a) the modal score

> The mode is the most common or most repeated result.

.... **(2 marks)**

(b) the median score

> The median is the score found in the middle position when all the scores are placed in order of size. If there is an even number of scores we find the half-way value between the two middle results.

.... **(2 marks)**

(c) the mean score

> The mean score is found by finding the sum of all scores, then dividing this result by the number of data.

(d) the range of scores

> The range of scores is found by subtracting the smallest score from the largest.

.... **(2 marks)**   .... **(2 marks)**

2 The annual salaries of 5 employees at a firm were given as the following.
£25 000, £31 500, £19 800, £64 000, £36 700

(a) Calculate the mean salary.

(£25 000 + £31 500 + ... + ... + ...) ÷ 5 = £............ **(2 marks)**

(b) A sixth employee joins the firm. She earns £19 800. What is the new mean?

.... **(2 marks)**

3 Erna wants to put numbers on five blank cards so that the mean is 10.8, the mode is 8, the median is 9 and the range is 8.

Work out the five numbers she must put on the cards.

☐ ☐ ☐ ☐ ☐

....   **(3 marks)**

Averages from tables 1

1 The table shows the number of countries outside the UK that students in a class have visited.

C	f	Cf
0	4	$0 \times 4 = \ldots\ldots\ldots$
1	7	
2	10	
3	6	
4	5	
$\Sigma(f)$		
		$\Sigma(Cf)$

> **C** means the number of countries visited.
> **f** means frequency – the number of students in each category.
> **Cf** means the number of visits.
> $\Sigma(f)$ means the sum of the frequencies.
> $\Sigma(Cf)$ means the sum of the numbers of visits for all categories.

(a) Complete the table. **(1 mark)**

(b) Work out the mode.

.... **(1 mark)**

(c) Work out the median.

> You can find the median by dividing $\Sigma(f)$ by 2. All you need do then is find which value of C this value links to.

.... **(2 marks)**

(d) Work out the mean.

> Use $\Sigma(Cf)$, representing all the visits in total, and divide it by $\Sigma(f)$ – the total number of students.

.... **(3 marks)**

(e) Work out the range.

.... **(2 marks)**

Guided

2 The table shows the number of visits to a doctor's surgery by people working in a small company, over the period of one year.

V	f	Vf
0	8	
1	5	
2	6	
3	4	
4	3	
5	1	
$\Sigma(f)$		$\Sigma(Vf)$

(a) Complete the table. **(1 mark)**

(b) Ramesh says that the mode of this data is 8. Explain why he is wrong.

....

(1 mark)

(c) Work out the correct mode, median and mean of this data.

Mode = ...

Median = ...

Mean = ... **(6 marks)**

PROBLEM
SOLVED!

(d) Kelly asks one more person. He visited the surgery 0 times last year. Kelly says this will not affect the median number of visits. Is she correct?

>
> You will need to use problem-solving skills throughout your exam – **be prepared!**

.... **(2 marks)**

121

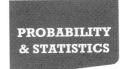
Averages from tables 2

PROBLEM SOLVED!

3 The table shows the number of hours spent on homework last week by pupils in a mathematics group.

Number of hours (h)	Frequency (f)	midpoint (x)	fx
$0 \leqslant h < 2$	1	1	$1 \times 1 = \ldots\ldots\ldots$
$2 \leqslant h < 4$	5	3	
$4 \leqslant h < 6$	6		
$6 \leqslant h < 8$	3		
$\Sigma(f)$			$\Sigma(fx)$

You will need to use problem-solving skills throughout your exam – **be prepared!**

(a) Complete the table.

> Because there is not one specific value but a class group instead, take the midpoint of the group to multiply the frequency.

(2 marks)

(b) Work out the modal class.

....

> The modal class is the most common value.

(1 mark)

(c) Which class contains the median?

....

> The median is the middle value.

(2 marks)

(d) Work out an estimate for the mean time spent doing homework.

$$\text{Estimated mean} = \frac{\Sigma fx}{\Sigma f}$$

$$= \frac{\ldots\ldots\ldots\ldots}{\ldots\ldots\ldots\ldots}$$

> To calculate an estimated mean, calculate the estimated total number of hours (Σfx) and divide by the total number of pupils (Σf).

(3 marks)

(e) Work out the range.

....

> The range is the difference between the largest and smallest values.

(2 marks)

Guided

4 A survey was taken in a Belgian school on how much pocket money students received each week. The table shows the results.

Money (€)	f		
$0 \leqslant € < 10$	4		
$10 \leqslant € < 20$	6		
$20 \leqslant € < 30$	8		
$30 \leqslant € < 40$	3		

(a) Complete the table.

(1 mark)

(b) Work out an estimate for the mean amount of pocket money received by these children each week.

....

(4 marks)

(c) Méganne forgot to add herself to the list. She now adds her weekly pocket money of €28.50 to the data. She says this will raise the mean and the median. Show why you either agree or disagree with her.

....

(2 marks)

Line graphs

1 The table records the annual number of visitors to a theme park from 2007 to 2016.

Year	Vistors
2007	2.4
2008	2.52
2009	2.65
2010	2.75
2011	2.65
2012	2.4
2013	2.5
2014	2.575
2015	1.925
2016	1.98

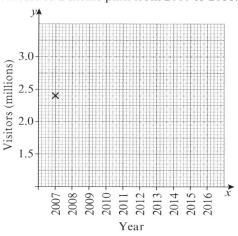

(a) Draw a time series line graph from the information given.

(3 marks)

(b) Describe the trend over the 10 year period.

> Use correct language like upward and downward when describing a trend.

(1 mark)

2 Kambiri counted the emails he received each day for one calendar month. The results are shown on the graph.

> Draw a frequency table to help you.

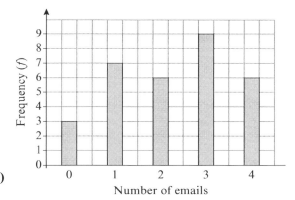

(a) There are 30 days in June. Explain how you know that Kambiri did not do his survey in that month.

.... **(3 marks)**

(b) What was the median number of emails?

> Find half the total frequency and count along the bars to see where that value falls.

(2 marks)

3 The vertical line graph shows the results of a survey of the shoe sizes of customers at a ladies' shoe shop.

(a) What is the modal shoe size?

.... **(1 mark)**

(b) Explain why the mean shoe size is unhelpful in this survey.

....

....

.... **(2 marks)**

123

Stem-and-leaf diagrams

1 Here are the results scored by 20 students in an end-of-term algebra test.
32, 41, 38, 29, 42, 50, 25, 19, 37, 30, 31, 42, 38, 39, 49, 19, 28, 44, 27, 11

(a) Complete the stem-and-leaf diagram to show this information.

> Remember to complete the key and give your diagram a title.

```
5 | 0
4 | 1  2  2
3 |
2 |
1 |
0 |
```

Key: 5 | 0

(b) Write down the modal score.

> Find the most common score.

.... **(1 mark)**

(c) Work out the median score.

> Order the scores and then find the middle one.

.... **(2 marks)**

(d) Work out the range.

> Subtract the smallest score from the largest score.

.... **(2 marks)**

(3 marks)

2 The number of passengers in one train carriage was recorded for one day on the journey between Guildford and Waterloo.
25, 41, 53, 49, 72, 91, 103, 115, 124, 105, 97, 86, 95, 92, 71, 68, 34, 61, 87, 97

(a) Draw a stem-and-leaf diagram to show this information.

(3 marks)

(b) Write down the modal passenger number.

.... **(1 mark)**

(c) Work out the median passenger number.

.... **(2 marks)**

(d) Work out the range.

.... **(2 marks)**

Key:

3 To the nearest pound, Marcella recorded customers' spending on costume hire at her party shop. She put the information on a stem-and-leaf diagram.

```
2 | x 7 7 9
3 | 5 6 6 9 9
4 | 0 0 1 2 4 4 5 8
5 | 1 1 1 3 7
6 | 0 2 2 4
7 | 0 0 1 2
```

Key : 3 | 5 = £35

(a) By looking at the leaf parts only, explain how you know that there is a modal value.

.... **(1 mark)**

(b) Write down the modal sale value.

.... **(1 mark)**

(c) Work out the median sale value.

.... **(2 marks)**

(d) The range of sales is £49. Find the value of x.

.... **(3 marks)**

Sampling

1 Dermot wanted to find out how much pocket money students in his class got each month. He did a small survey among his friends and got the following results.
£30, £25, £40, £20, £50, £40, £35, £120

(a) Use Dermot's data to estimate the mean amount of pocket money for the whole class.

.... **(2 marks)**

(b) Estimate the modal value for the whole class.

.... **(1 mark)**

(c) How reliable do you think your estimate is? Explain your answer.

> Think about the size of the sample and the size of the class, and about how Dermot chose the people for his survey.

....

.... **(2 marks)**

(d) How could Dermot get a more reliable estimate?

....

.... **(2 marks)**

2 A sports scientist carried out an experiment at an athletics club. She measured the time runners took to run 100 m, and recorded the length of each runner's lower leg, to the knee. She plotted the results on a scatter graph and drew a line of best fit through the points.

(a) Use the line of best fit to estimate the running time of an athlete with a lower leg length of 47 cm.

.... **(1 mark)**

(b) Use the line of best fit to estimate the lower leg length of a runner who completed the 100 m in 14 seconds.

.... **(1 mark)**

(c) Write down a suggestion for how the estimates could be improved.
.... **(1 mark)**

(d) Angelina says the line of best fit could not be made any longer. Explain why she is correct.

....

....

.... **(3 marks)**

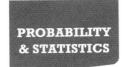

Had a go ☐ Nearly there ☐ Nailed it! ☐

Stratified sampling

1 A large company has a sales force of 700, based in different regions of the mainland UK. The table shows a stratified sample of where 140 of them are based.

South-west	Midlands	South-east	North-west	North-east
18	33	43	17	29

(a) Work out how many of the company's sales force is based in the north-west.

Number based in north-west $= \left(\dfrac{17}{140}\right) \times$ Total sales force

$= \dfrac{17 \times 700}{140} = \dfrac{17 \times 10}{2} = \ldots\ldots\ldots\ldots$

Number of sales force based in north-west $= \ldots\ldots\ldots\ldots$ **(2 marks)**

(b) Three years previously, the same stratified sample gave identical results but the company sales force was only 630. Work out how many of the sales force were based in the midlands three years ago.

$\ldots\ldots\ldots\ldots\ldots\ldots\ldots\ldots\ldots\ldots$ **(2 marks)**

2 Hendrik is going to do a survey on the creative skills of workers in a large manufacturing firm. He has selected a sample of 75 workers, stratified by gender and type of job. The table shows this information.

There are 280 male workers altogether.

	Male	Female
Office	15	24
Factory	20	16

(a) Work out the total number of female workers.

Number of females in sample $= \ldots\ldots\ldots + \ldots\ldots\ldots = \ldots\ldots\ldots$

Number of males in sample $= \ldots\ldots\ldots + \ldots\ldots\ldots = \ldots\ldots\ldots$

$\dfrac{females}{males} = \dfrac{\ldots\ldots\ldots}{\ldots\ldots\ldots}$

> Multiply this by the total number of male workers.

$= \dfrac{\ldots\ldots\ldots}{\ldots\ldots\ldots} \times 280 = \dfrac{\ldots\ldots\ldots}{\ldots\ldots\ldots} = \ldots\ldots\ldots\ldots\ldots\ldots$ **(3 marks)**

(b) Calculate the number of male office workers in the firm.

Number of male office workers in sample $= \ldots\ldots\ldots\ldots\ldots\ldots$

Total number of male workers in sample $= \ldots\ldots\ldots\ldots\ldots\ldots$

In the sample, $\dfrac{\ldots\ldots\ldots}{\ldots\ldots\ldots} = \dfrac{\ldots\ldots\ldots}{\ldots\ldots\ldots}$ of the male workers are office workers.

Total number of male office workers $= \dfrac{\ldots\ldots\ldots}{\ldots\ldots\ldots} \times 280 = \dfrac{\ldots\ldots\ldots}{\ldots\ldots\ldots} = \ldots\ldots\ldots$ **(3 marks)**

3 The table shows information about the number of people in a Welsh village who use a computer for more than 4 hours a day.
Morwenna is going to take a sample of 40 people, stratified by gender and age.
How many females over 45 should be in the sample?

	Age		
	Under 18	**18 to 45**	**Over 45**
Male	132	246	107
Female	126	186	123

$\ldots\ldots\ldots\ldots\ldots\ldots$ **(3 marks)**

Comparing data

1 The following table shows a teacher's analysis of how students performed in two language exams, both marked out of 80. The Head of Department says the table shows that students do better in German than in Italian. Give two pieces of evidence that show this is true.

	Mean	Range
German	60	12
Italian	54	13

The mean score in German is ....

The range of test scores in .... is less than that in .... **(2 marks)**

2 The table shows the results of two Mathematics tests sat by a group of Year 10 students, both marked out of 100.

Compare the test scores in Algebra and Statistics.

	Mean	Range
Algebra	63	8
Statistics	64	16

....

.... **(2 marks)**

3 Mrs Wright made a bar chart of absences in her class over a four-week period.

(a) Work out the median of this data and explain what it tells you.

> Work out the middle value of the ordered data.

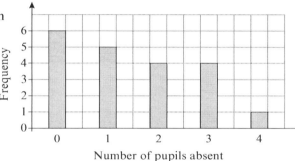

Number of pupils absent

.... **(2 marks)**

(b) Work out the range.   Work out the difference between the smallest and the largest data values.

.... **(1 mark)**

Mr Patel did the same for absences in his class. The results had a range of 5 and median of 3.

(c) Compare the number of absences for each class.

....

.... **(2 marks)**

4 The exam marks in English for two schools were collected and 16 students from each school were randomly sampled. The results are shown on a back-to-back stem-and-leaf diagram.

Beauregard High   McDowall Academy

```
                 4 | 9
            6 1 | 5 | 2 4 5 5 6
 9 9 8 7 7 5 4 4 2 1 | 6 | 1 3 3 6
           3 1 0 | 7 | 4 5 7
               0 | 8 | 4 4
                 9 | 2
```

Key: 1 | 5 = 51   Key: 7 | 3 = 73

Compare the data from the two schools. Comment on which group of students performed better in English.

....

....

....

.... **(4 marks)**

127

Had a go ☐ **Nearly there** ☐ **Nailed it!** ☐

Probability 1

1 On the probability scale, mark with a cross (x) the probability that

Remember, the probability scale goes between 0 and 1, where 0 represents an impossibility and 1 represents a certainty.

(a) Tuesday will follow Monday.

0   1

(1 mark)

(b) a random person has a birthday in November.

0   1

(1 mark)

(c) a dice roll scores lower than 3.

0   1

'Lower than 3' is not the same as '3 or lower'.

(1 mark)

(d) two fair coins both land on heads.

Probability of heads $= \frac{1}{2}$

$\frac{1}{2} \times \frac{1}{2} = \frac{1}{4}$

Each coin is independent and you want heads on both. You multiply the probabilities to find how likely this is.

0   1

(1 mark)

2 Harry takes all 12 picture cards (kings, queens and jacks) from a pack of playing cards and lays them face down on a table. He picks a card.
Write down the theoretical probability that Harry chooses

(a) a king.

List all 12 possible cards. How many are kings?

.... **(1 mark)**

(b) a red queen.

Write down how many twelfths the red queens make up.

.... **(1 mark)**

(c) either a black king or a red jack.

$$\frac{\dots + \dots}{12} = \frac{\dots}{\dots}$$

$= \dots$ **(1 mark)**

(d) an ace.

.... **(1 mark)**

3 From the following list, which term best describes each event?

certain   **likely**   **even chance**   **unlikely**   **impossible**

Diamonds   Clubs   Hearts   Spades

(a) Ten playing cards are drawn from a shuffled pack of 52. At least one is a heart or a diamond.

There are 13 cards in each suit.

.... **(1 mark)**

(b) It will get dark tonight in Edinburgh.

.... **(1 mark)**

(c) A banknote chosen at random from a person's wallet will be a £7 note.

.... **(1 mark)**

Probability 2

4 Students in one class use three different modes of transport to school. The table shows the probabilities of a randomly chosen student using two of those modes of transport.

Modes of transport	Walking	Bus	Car
Probability	0.52	....	0.16

A student is chosen at random. Work out the probability that this student

(a) walks to school or arrives by car.

0.52 + ... = ...   **(1 mark)**

(b) arrives at school by bus.   │ The probability of all possible events must equal 1. │

1 − .... − .... = ...   **(2 marks)**

5 The diagram shows a 20-sided spinner. Complete the probability table for spinning each number.

│ Write the probabilities as fractions, then turn them into decimals. │

Number	0	1	2	3	4	5	6
Probability	......	0.25	......	......	......	0.1	......

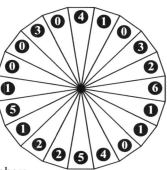

(7 marks)

6 There are four house teams in a chess tournament. Some team members are Year 10s and some are Year 11s. The probabilities of a win are shown. In team Wilberforce the probability of a win is the same for Year 10s as for Year 11s.

		Fry	Howard	Pankhurst	Wilberforce
Probability	Year 10s	0.16	0.04	0.2	....
	Year 11s	0.06	0.18	0.12	....

(a) A student is chosen at random to give a talk about the chess tournament. Calculate the probability that a winning Year 11 student is chosen.

....   **(2 marks)**

(b) A winning student is chosen at random to give a talk to younger students about Chess Club. Calculate the probability that they are from team Howard.

....   **(2 marks)**

(c) A winning Year 11 student is chosen to write about the tournament in the school newsletter. Calculate the probability that the student will be from team Fry.

....   **(2 marks)**

(d) Calculate the probability that a winning player is a Year 10 from Wilberforce or a Year 11 from Howard.

....   **(2 marks)**

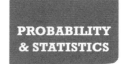
Had a go ☐ **Nearly there** ☐ **Nailed it!** ☐

Relative frequency

1 The daily sales of fiction titles by a bookshop were as follows.

Genre	Romance	Sci-Fi	Thriller	Humour	Historical
Sales	18	17	21	19	5

Calculate, to 2 significant figures, the probability that a sale chosen at random was

(a) a historical novel.

Probability {historical} = $\dfrac{5}{\text{......}}$ = ...

> Make a fraction of the historical genre divided by all possible genres.

(2 marks)

(b) a romance, a thriller or a humorous work.

19 + ... + ... = ...

$\dfrac{..........}{.........}$ = ...

(2 marks)

2 The table shows the total scores when Marie-Louise fired three arrows 20 times.

Score	0–15	20–40	45–75	80–115	120–150
Frequency	1	5	7	5	2

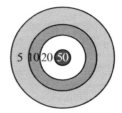

Estimate the probability that the score from the arrow will be

(a) 45–75.

...... out of 20 = $\dfrac{......}{20}$ = ... **(1 mark)**

(b) more than 75.

> You will need to combine two categories.

.... **(2 marks)**

(c) no more than 15.

.... **(2 marks)**

3 Kerry is the opening bat for her cricket team and has recorded her run scores for the last 20 matches.

Guided

(a) The next match she plays is the last in the season. Calculate the estimated probability that in this match she scores 60 or fewer runs.

.... **(2 marks)**

Runs (r)	Frequency
$0 < r \leqslant 20$	2
$20 < r \leqslant 40$	5
$40 < r \leqslant 60$	9
$60 < r \leqslant 80$	3
$80 < r \leqslant 100$	1

(b) Comment on the accuracy of your estimate in part (a).

....

....

(1 mark)

(c) Jonah plays cricket to the same standard as Kerry, but he played 35 matches this season. Calculate an estimate for the number of matches in which he scored over 60 runs.

.... **(1 mark)**

Frequency and outcomes

1 Dirk is going to buy an ice cream. He can choose a plain cone or a chocolate cone with a choice of vanilla, chocolate or peach ice cream. He likes all the choices equally!
Work out the probability that Dirk chooses a plain cone and peach ice cream. Write it as a fraction.

| How many equal outcomes are there? |

Flavour

Cone

.... Plain .... Vanilla
.... Chocolate
.... Peach

.... Chocolate .... Vanilla
.... Chocolate
.... Peach

P(plain cone with peach ice-cream) = P(Plain cone) × P(peach ice cream)

$$= \frac{\cdots}{\cdots} \times \frac{\cdots}{\cdots} = \frac{\cdots}{\cdots}$$ **(2 marks)**

2 In a survey of university graduates, 75 people are asked about their employment. There are 42 science graduates, 8 of whom are unemployed. The rest are humanities graduates. In total, 57 people are employed.

(a) Complete the frequency tree.

| First work out the number of humanities graduates.
Then work out the number of employed humanities graduates. |

(2 marks)

(b) What is the probability that a person chosen at random is employed?

.... **(1 mark)**

| To turn the numbers into probabilities, write them as fractions of the people surveyed, with 75 in the denominator. |

(c) A humanities graduate is chosen at random. What is the probability that this person is unemployed?

.... **(2 marks)**

3 Two packs of playing cards, both with 52 cards each, are opened and mixed together. Then 60 cards are randomly chosen to make up a new pack. The probability of randomly choosing a diamond from this new pack is $\frac{1}{5}$, and the probability of randomly choosing a red card is $\frac{3}{10}$.

> **Guided**

(a) What is the probability of choosing a black card at random?

.... **(2 marks)**

(b) How many hearts are in the new pack of 60?

.... **(2 marks)**

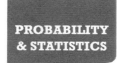

Had a go ☐ **Nearly there** ☐ **Nailed it!** ☐

Venn diagrams

1 The diagrams show what a group of 42 conference guests ate for breakfast on three consecutive days.

> A Venn diagram is a way of grouping different items. These groups are known as sets.

Monday

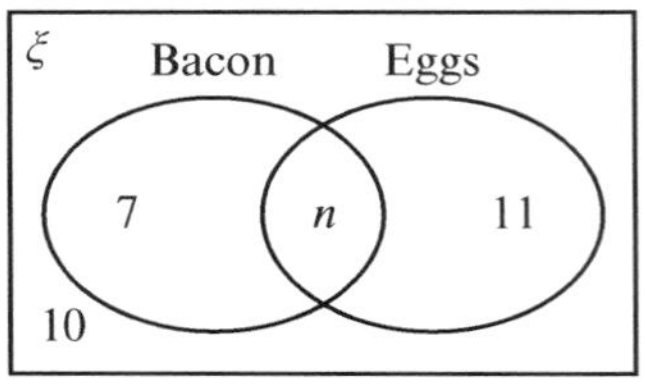

ξ Bacon   Eggs

7   n 11

10

(a) What does n represent in the Venn diagram for Monday?

.... **(1 mark)**

(b) Work out the value of n in the diagram for Monday.

> Everything in the universal set (ξ) totals 42

42 − 10 − .... − .... − .... = ...

n = ...   **(2 marks)**

(c) This diagram shows two menu choices for Tuesday. The people who had only cereal are recorded in the area labelled n. Work out the value of n.

Tuesday

ξ Sausage   Cereal

6   3 n

8

> The total is still 42. Notice that n means something different this time.

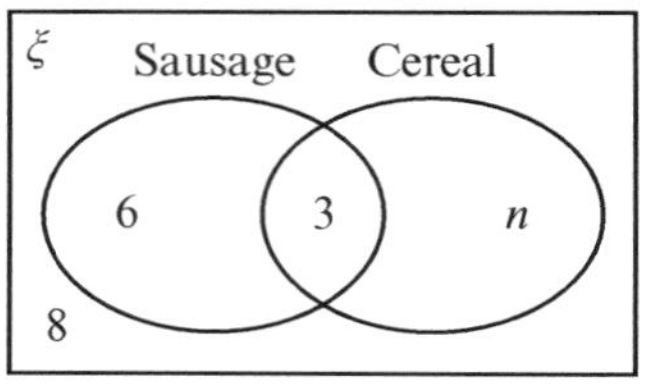

....   **(2 marks)**

(d) This diagram shows two menu items for Wednesday. Work out how many people had neither muesli nor toast.

Wednesday

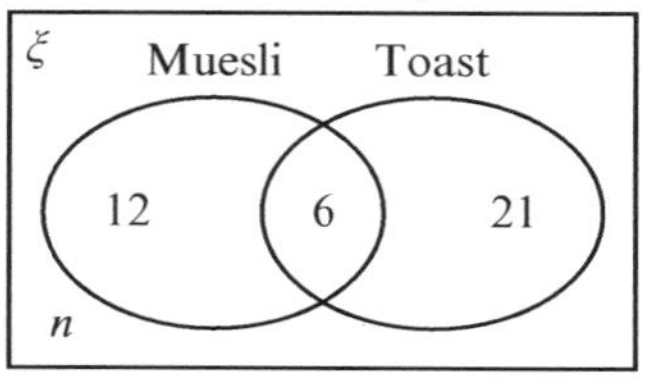

ξ Muesli   Toast

12   6   21

n

n = 42 − ....   **(2 marks)**

2 In a class of 30 students, 8 have pet dogs only, 5 have pet cats only and 11 have both.

(a) Draw a Venn diagram to show this information.

ξ

(2 marks)

(b) What is the probability that a student chosen at random does not have either of these pets?

....   **(2 marks)**

(c) What is the probability that a student chosen at random has only a pet dog or no pet at all?

....   **(2 marks)**

Independent events

1 Karen has a fair 6-sided spinner. She spins it and takes a note of the result. She then spins it again.

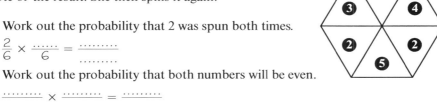

(a) Work out the probability that 2 was spun both times.

$\dfrac{2}{6} \times \dfrac{\text{........}}{6} = \dfrac{\text{..........}}{\text{........}}$ **(2 marks)**

(b) Work out the probability that both numbers will be even.

$\dfrac{\text{..........}}{\text{........}} \times \dfrac{\text{..........}}{\text{........}} = \dfrac{\text{..........}}{\text{........}}$ **(2 marks)**

(c) Work out the probability that one number is odd and the other is 2.

$\dfrac{\text{..........}}{\text{........}} \times \dfrac{\text{..........}}{\text{........}} + \dfrac{\text{..........}}{\text{........}} \times \dfrac{\text{..........}}{\text{........}} = \dfrac{\text{..........}}{\text{........}}$ **(3 marks)**

2 A bag contains 7 German euro coins and 13 Greek euro coins. All euro coins are the same shape and weight. A coin is taken out and replaced, then another coin is taken out.

(a) Complete the sample space diagram and tree diagram.

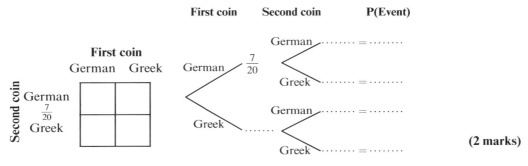

(b) What is the probability that a German coin is **not** chosen both times?

....   **(2 marks)**

3 Maya and Gianluca are throwing darts. The probability that Maya hits the bullseye (the centre of the target) is 0.3, and the probability that Gianluca hits the bullseye is 0.2.

(a) Continue and complete the tree diagram.

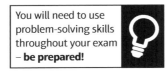

You will need to use problem-solving skills throughout your exam – **be prepared!**

Maia   Gianluca

Bullseye ....

No bullseye ....

(2 marks)

(b) What is the probability that only one of them hits the bullseye?

....   **(3 marks)**

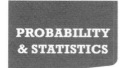

Had a go ☐   Nearly there ☐   Nailed it! ☐

Problem-solving practice 1

1 60 people were asked their favourite type of restaurant. The two-way table shows some of the results.

(a) Complete the table.   **(2 marks)**

(b) What is the probability that a person chosen at random from those who preferred Indian food was male?

	Chinese	Indian	Italian	Total
Male	7		14	
Female			8	27
Total	16			60

.... **(2 marks)**

2 A school offers each Year 11 student a choice of 1 of 5 after-school classes as shown in the table.

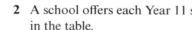

	Badminton	Pottery	Spanish	Guitar	Chess
Probability			0.15	0.4	0.05

A randomly chosen student is three times more likely to do badminton than pottery.

(a) Complete the table.

(3 marks)

(b) What is the probability that a randomly chosen student does not do pottery?

.... **(2 marks)**

(c) There are 85 students who have signed up for an after-school class. Estimate the number that are doing badminton.

.... **(2 marks)**

3 A group of 15 students from a mathematics class and another 15 from a woodwork class were asked to draw a line using a straight edge, but not a ruler, as close as possible to 7 cm long. The lines were then measured to 1 decimal place and recorded on the stem-and-leaf diagram.

Woodwork										Mathematics			
							1	8					
						2	1	7	1	3	3	6	
5	4	4	3	1	1	1	0	6	0	0	1	2	
				9	9	8	7	5	5	5	7	9	
								4	8	9	9		

Key: 1 | 8 = 8.1 cm   Key: 5 | 7 = 5.7 cm

Compare the results from both classes and comment on which class has better estimation skills.

.... **(3 marks)**

Problem-solving practice 2

4 Kirsty did a survey of 90 train passengers at a station. She asked them if they were happy, unhappy or undecided about the train service. Complete the table from information marked on the pie chart.

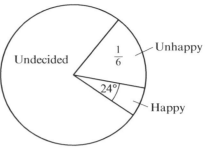

Response	Frequency	Angle
Happy	6	24°
Unhappy		
Undecided	69	
Total	90	

(4 marks)

5 Angela and Andy are about to take their driving tests. The probability of Angela passing is 0.85 and the probability of Andy failing is 0.35.

(a) Complete the probability tree diagram.

   Angela   Andy

   Pass ⟨ · · · · · · · · · · ·

   ⟨

   Fail ⟨ · · · · · · · · · · ·   **(3 marks)**

(b) What is the probability that just one of them will fail?

   ·   **(3 marks)**

6 A group of 42 students at a college in Wolverhampton were asked if they had ever visited Brighton or London. 14 had been to London, 5 to Brighton and 12 to both London and Brighton.

(a) Draw a Venn diagram to show this information.

   ξ

   (2 marks)

(b) What is the probability that a student chosen at random had not visited either London or Brighton?

   ·   **(2 marks)**

(c) What is the probability that a student chosen at random had visited both places or neither?

   ·   **(2 marks)**

Paper 1

Practice exam paper

Foundation Tier
Time: 1 hour 30 minutes
Calculators must not be used
Diagrams are **NOT** accurately drawn, unless otherwise indicated.
You must **show all your working out.**

1. Karl wrote down the heights above/below sea level of 6 international cities. He recorded his findings in a table.

City	Metres above sea level
Dublin	8
Amsterdam	–2
Cusco (Peru)	3399
Djibuti	0
Johannesburg	1753
Baku	–28

Write down

(a) the city at the lowest height.

.... **(1 mark)**

(b) the mean height above sea level of the 6 cities.

.... **(1 mark)**

(c) the median height above/below sea level of the 6 cities.

.... **(1 mark)**

(d) the range of heights.

.... **(1 mark)**

2. The owners of a guitar shop supplied their accountant with the following information about the four-month pre-Christmas period. Some figures were estimates (shown in italic on the left).

Month	Costs (£)	Sales (£)	Profit (sales – costs)
September	4134	*6800*	2659
October	4478	6891	*2400*
November	*4300*	7543	3225
December	4592	*7300*	2684
Total	*18000*	*29000*	

(a) Calculate exact figures for the estimates and enter them in the table.

.... **(3 marks)**

(b) Write down the value of guitar sales in December.

.... **(1 mark)**

(c) Write down the lowest monthly cost of running the shop during this period, to 2 s.f.

.... **(1 mark)**

(d) Calculate the range of guitar sales during this period.

.... **(1 mark)**

(e) Calculate the total profit for the four-month pre-Christmas period.

.... **(2 marks)**

(f) Calculate the mean profit per month during this period.

.... **(2 marks)**

3. A 12-sided fair spinner is labelled as in the diagram.

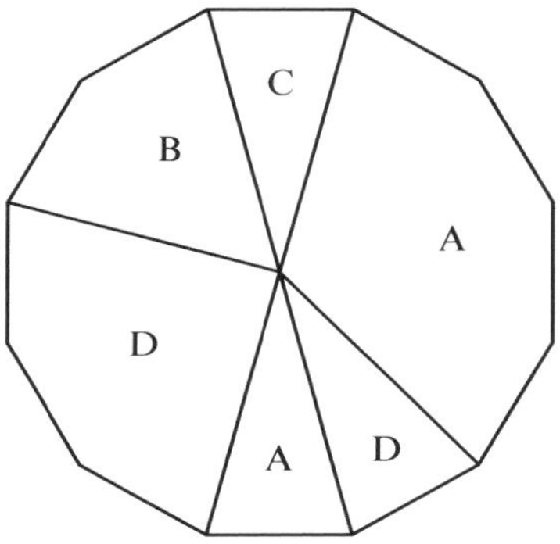

(a) If the spinner is spun once

(i) explain why A is the most likely outcome.

.... **(1 mark)**

(ii) calculate the probability of the outcome being B or D.

.... **(2 marks)**

(b) If the spinner is spun twice what is the probability of getting an A both times?

.... **(2 marks)**

4. Three similar triangles *ABC*, *CDE* and *EFG* are labelled on a map and joined together, forming a straight line *ACEG*.

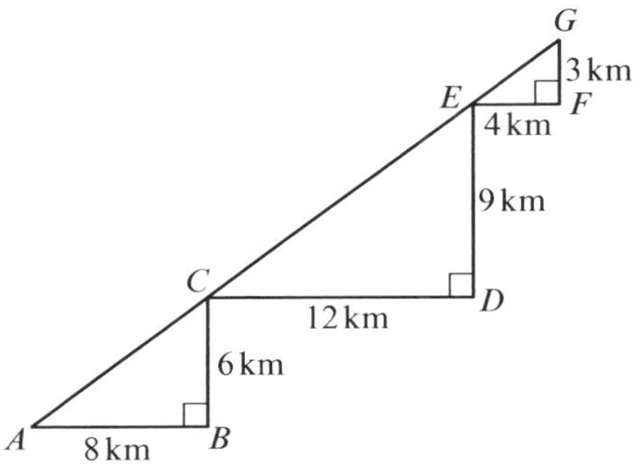

(a) Work out the shortest distance from *A* to *G*.

.... **(2 marks)**

(b) Tristan cycles from *A* to *B* then to *C*. Bonnie cycles from *C* to *E*. Who has the shorter journey and by how many kilometres?

.... **(2 marks)**

(c) If another triangle with a horizontal length of 25 km and a vertical length of 20 km was drawn, would this be similar to the three triangles drawn? Explain your answer.

....

....

(2 marks)

5. A dual bar chart shows the results of a survey of male and female shoppers when asked where they did most of their food shopping.

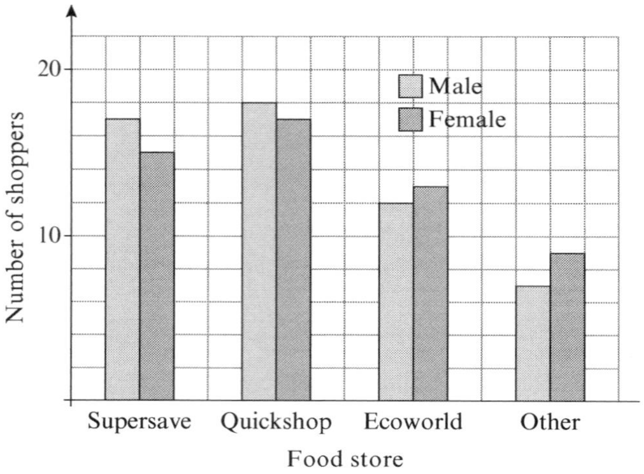

Complete the following table showing the information from the bar chart.

(2 marks)

6. Simplify the following expressions.

(a) $x + 2x + 5x - 7x + x$

.... **(1 mark)**

(b) $4a + 4a + 4a$

.... **(1 mark)**

(c) $9x + 2y - 4x + 5y$

.... **(1 mark)**

(d) $2x^2 - 2 - 5x + 4x^2 + x$

.... **(1 mark)**

7. George correctly thinks that 18×22 is the same as $20^2 - 2^2$ and he says he used the difference of two squares to solve it.
Solve the following using the same method as George.

(a) 33×27

.... **(1 mark)**

(b) 10.5×9.5

.... **(1 mark)**

8. Write the following fractions in order of size starting with the smallest.

$$\frac{3}{8} \quad \frac{1}{3} \quad \frac{2}{5} \quad \frac{3}{7} \quad \frac{4}{9}$$

.... **(2 marks)**

9.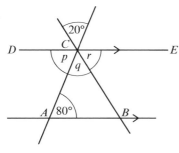

(a) What type of triangle is ABC?

.... **(1 mark)**

(b) DCE is a straight line, therefore, $p + q + r = 180°$. Label all the angles using the facts about alternate, vertically opposite and corresponding angles, to show that the interior angles of a triangle also sum to $180°$.

.... **(3 marks)**

10. The length of Loch Ness is 36 km. Derek's map shows this distance as 18 cm. Work out the scale of the map.

.... **(2 marks)**

11. Factorise the following expressions.

(a) $12x - 8y + 16$

.... **(1 mark)**

(b) $5x^3 - 25x^2 + 15x$

.... **(1 mark)**

12. On a recent trip to Milan, Dara brought back €28. The exchange rate at his bank in England was £1 = €1.1.

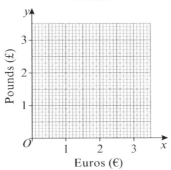

Conversion rate:
£ and €

(a) Draw a graph of this conversion rate.   **(1 mark)**

(b) Work out from your graph how much Dara should expect for his euros.

.... **(1 mark)**

(c) Dara's friend Siobhan is going out to Paris on the same day that Dara gets back. She needs to change £560 into euros. Use the graph to estimate how much she will get.

.... **(2 marks)**

13. A 7.25 kg sack of flour is emptied into containers that hold 37.5 g of flour. Calculate the number of containers needed.

.... **(2 marks)**

14. Marie bought a new car for £17 850. At the end of the second year after she bought it, the car had depreciated by 66%. By the end of the third year it had a value of only £4998.

(a) Calculate the value of the car at the end of the second year.

.... **(3 marks)**

(b) Calculate the total percentage depreciation of the car by the end of the third year.

.... **(3 marks)**

15. Fifty cruise ship passengers were asked which units of currency they had in their wallets. 42 had euros, 11 had euros and US dollars, 19 had both pounds and euros, 2 had all three, 0 had both pounds and US dollars. The number of passengers with none of these currencies was equal to the number who had only pounds plus the number who had only US dollars. There were three times more passengers with just US dollars than with just pounds.

(a) Complete the Venn diagram to show this information.   **(3 marks)**

(b) What was the probability that a passenger chosen at random did not have pounds in their wallet?

.... **(2 marks)**

16. The shape shown has a perimeter is 56 cm.
Calculate the area.

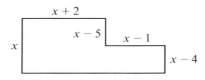

.... **(4 marks)**

17. Four towns, A, B, C and D, lie on a straight line. A to B is 7.5 km. B to C is $\frac{1}{4}$ the distance of A to D.
C to D is $\frac{2}{5}$ of the total distance.

What is the distance of A to D?

.... **(3 marks)**

18. Here are the first seven terms in a sequence. The terms p, q and r have not yet been calculated.

-6 -2 2 6 p q r

(a) What is the value of term r?

.... **(2 marks)**

(b) Explain why 260 cannot be a term in this sequence.

. . . . **(1 mark)**

19. Sally and Dan are 3.4×10^6 cm away from each other.

(a) How far is this as an ordinary number? Write your answer in a reasonable standard unit of measurement.

.... **(1 mark)**

(b) Which is the longer distance, 5.23×10^5 mm or 5.21×10^{-2} km?

.... **(2 marks)**

20. Calculate angle a in the irregular pentagon.

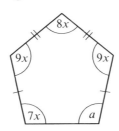

.... **(2 marks)**

141

21. (a) Draw the graphs of $2x - y = 8$ and $x + 2y = -1$ on the same set of axes by first completing the tables.

$2x - y = 8$

x	0	2	4
y			

$x + 2y = -1$

x	−3	1	3
y			

(2 marks)

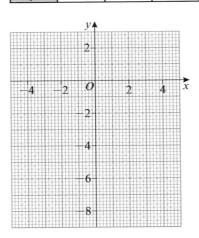

(b) What are the only values of both x and y that are true for both equations? Give your answer as a pair of coordinates.

.... **(1 mark)**

(2 marks)

22. The two spinners below are each spun and the results are multiplied to give a score of A.

Complete the tree diagram to find the probability that $A = 6$.

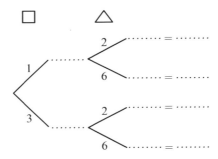

.... **(2 marks)**

Practice exam papers for Paper 2 and Paper 3 are available to download free from the Pearson website. Scan this QR code or visit http://activetea.ch/1MpXwRb.

Answers

NUMBER

1. Place value

1. (a) 25 302
   (b) twelve thousand, three hundred and seventeen
   (c) 7000

2.

10 000s	1000s	100s	10s	1s
1	9	0	6	1

3. (a) 199, 209, 211, 219, 251
   (b) 3099, 3118, 3711, 3796, 3910

4. (a) 29 738 km, 31 787 km, 31 809 km, 31 851 km, 32 001 km
   (b) 5.09 m, 5.11 m, 5.75 m, 5.8 m, 5.92 m

5. £6.70 and £6.35

6. 5

2. Negative numbers

1. (a) $-12, -5, 0, 3, 8$
   (b) (i) -9 (ii) 8 (iii) -30 (iv) -22

2. (a) -22 (b) -7
   (c) 49 (d) 16

3. (a) (i) $-2°$ (ii) $-9°$
   (iii) $-16°$ (b) $22°$

4. (a) 246 m (b) 130.5 m

3. Rounding numbers

1. (a) 11 300 (b) 13 000 (c) 22 000

2. (a) 0.003 (b) 0.003 27 (c) 0.0033

3. (a) 400 000 (b) 360 000 (c) 362 000

4. (a) 10.9 (b) 11 (c) 2

5. By giving the figure to 3 s.f. he has rounded the hundreds to the nearest 1000 so he is correct.

4. Adding and subtracting

1. (a) 950 (b) 547

2. (a) 9958 (b) 6533

3. £1.86

4. 41

5. The total is £10.23, which means Manny is incorrect.

5. Multiplying and dividing

1. (a) 1470 (b) 73

2. (a) 96 (b) 136

3. 702

4. (a) 4800 kg (b) 320 kg

6. Decimals and place value

1. (a) 4 hundredths
   (b) 2 hundredths
   (c) 3 thousandths

2. 4.9, 6.7, 7.6, 8.2, 8.3

3. 1.6, 1.532, 1.53, 1.504, 1.499

4. 0.019, 0.08, 0.788, 0.8, 0.81

5. (a) 111.6 (b) 1.116 (c) 111.6 (d) 93

6. Either 0.003 99 **or** 2100 **or** 190

7. Operations on decimals

1. (a) 12.33 (b) 53.2 (c) 58.92
   (d) 0.0156 (e) 10.2 (f) 41

2. £201.60   3. £15.20   4. £1192.00

8. Squares, cubes and roots

1. (a) 36 (b) 27 (c) 6
   (d) 12 (e) 3 (f) 4
   (g) 5 (h) -2 (i) -3

2. 3

3. (a) 8 (b) 29 (can also accept 2)
   (c) 64

4. It is true because 4 consecutive square numbers will consist of 2 odd and 2 even numbers which when added always gives an even sum.

9. Indices

1. (a) 5^3 (b) 5^6

2. (a) 3^7 (b) 4^8 (c) 5^9
   (d) 3^3 (e) 7^6 (f) a^3

3. (a) 4^{-1} (b) 4^{-3}

4. (a) 2^2 (b) 5^3 (c) 8^4 (d) 3^6

5. (a) 1 (b) $\frac{1}{25}$ (c) $\frac{1}{8}$ (d) $\frac{27}{8}$

6. 3

10. Estimation

1. (a) 9000 (b) 60 (c) 400

2. 40   3. 1   4. 50   5. 500   6. 24

7. (a) 100 cm³
   (b) Estimate based on $10 \times 40 \div 0.25$ will make it larger than real volume as first two numbers are higher estimations.

11. Factors, multiples and primes

1. (a) {1, 2, 3, 4, 6, 8, 12, 24}
   (b) 48, 56, 64, 72, 80, 88

2. 1, 7, 42, 70

3. 2, 3, 5, 7, 11, 13

4. (a) 1p, 2p, 4p, 5p, 10p, 20p, 25p, 50p, 100p
   (b) 9

5. (a) 12, 15, 20 (b) 47

6. (a) $2^2 \times 3 \times 7$ (b) $2^3 \times 5$ (c) $2 \times 3 \times 7$ (d) $2^3 \times 5^2$

12. HCF and LCM

1. (a) 15 (b) 60

2. $56 = 2^3 \times 7$, $196 = 2^2 \times 7^2$

3.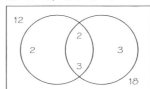

   LCM (12, 18) = 36

4. (a) 2 (b) 132

13. Fractions

1. $\frac{7}{12}$

2. (a) $\frac{1}{3}$ (b) $\frac{2}{5}$ (c) 1 (d) $\frac{13}{20}$

3. (a) $\frac{3}{5}$ (b) $\frac{2}{5}$

4. (a) 60 km (b) £16 (c) 27 kg

5. £100.80

14. Operations on fractions

1. (a) $\frac{1}{2}$ (b) $\frac{7}{12}$ (c) $\frac{17}{35}$ (d) $\frac{1}{36}$

2. (a) $\frac{1}{8}$ (b) $\frac{8}{15}$ (c) $\frac{2}{3}$ (d) $\frac{2}{3}$

3. $\frac{7}{20}$

4. (a) $\frac{13}{72}$ (b) £3.25

15. Mixed numbers

1. (a) $7\frac{19}{30}$ (b) $2\frac{19}{21}$

2. (a) $3\frac{1}{5}$ (b) $1\frac{5}{6}$

3. (a) $11\frac{5}{9}$ (b) $3\frac{1}{3}$

4. $9\frac{9}{28}$ 5. 52 km   6. $59\frac{4}{5}$ miles

16. Calculator and number skills

1. (a) 9 (b) 20.5 (c) 10 (d) $\sqrt{\frac{4}{3}}$

2. (a) $\frac{382}{21}$ (b) 6 (c) 1

3. 2

4 7

5 (a) 2.078 353 492   (b) 2.1

17. Standard form 1

1 (a) 3.5×10^3 (b) 0.0121   (c) 3 200
2 (a) 6×10^2 (b) 5.7×10^2 (c) 4.003×10^3
     (d) 5.11×10 (e) 6.113×10
3 (a) 3.5×10^{-1} (b) 1.82×10^{-3} (c) 4.87×10^4
4 (a) 5.842×10^6 (b) 5.85×10^6 to 3 s.f.
     (c) 5.84×10^6
5 21 300 mph to 3 s.f.

18. Standard form 2

6 (a) 2.8×10^3 (b) 5×10^7
7 (a) 1.82×10^3 (b) 5.07×10^6
8 (a) 6.4×10^3 (b) 2.24×10
     (c) $3.199\,986 \times 10^3$ (d) $6.857\,143 \times 10^5$
9 Earth 13.4 times greater surface area
10 133 000

19. Counting strategies

1 (P2, P4, P6) (Q2, Q4, Q6) (R2, R4, R6)
2

	Touring (T)	Mountain(M)	Fold-up (F)
Orange (O)	TO	MO	FO
Jade (J)	TJ	MJ	FJ
Silver (S)	TS	MS	FS

3

	10p	20p	50p	£1
£1	£1.10	£1.20	£1.50	
50p	£0.60	£0.70		
20p	£0.30			
10p				

4 456, 465, 546, 564, 645, 654
5 10 tracks (PG, PT, PF, PM, TG, TF, TM, FG, FM, MG)

20. Problem-solving practice 1

1 (a) $3 + 5$ (b) $5 + 7$
     (c) $7 + 13$ (d) $13 + 23$
2 £4.51   **3** 9 friends, £0.70 change   **4** 172
5 $\frac{3}{5} = \frac{3 \times 7}{5 \times 7} = \frac{21}{35}$ $\frac{5}{7} = \frac{5 \times 5}{7 \times 5} = \frac{25}{35}$ so $\frac{5}{7}$ is larger
6 813

21. Problem-solving practice 2

7 (a) $\frac{1}{3}$ (b) 60°

8 (a) 26   (b) £221

9 $28\frac{1}{3}$ cm

10 Anne = 4 circuits, Margaret = 3 circuits

11 $31\frac{1}{9}$ m/s   **12** $n = \frac{3}{2}$

ALGEBRA

22. Collecting like terms

1 (a) equation   (b) expression   (c) formula
     (d) equation   (e) formula
2 (a) $6n - 2$ (b) $8ab$
3 (a) $8p + 3q$ (b) $10pqr$
4 (a) $7x$ (b) $10y^2$
     (c) $8x^2 + 2x$ (d) $30 - 3n$
5 (a) $3a^2$ (b) $3a^2 - 2a + b$ (c) $3x^2 - 2x + 5$

23. Simplifying expressions

1 (a) a^3 (b) $7ab$ (c) $2pqr$
2 (a) $2b^3$ (b) $9ab$ (c) $20q$ (d) $22gh$
3 (a) $6y^2$ (b) $18mn$ (c) $3a$ (d) $15x$
4 (a) $35ac$ (b) $30u^3$ (c) $6g$ (d) $3a$
5 (a) $20pqr$ (b) $6f$
6 $2x \times 2x$

24. Algebraic indices

1 (a) y^6 (b) n^2 (c) f^6
2 (a) a^6 (b) q^7 (c) b^4 (d) c^8

3 (a) b^{20} (b) $8y^6$ (c) $18a^7$
     (d) $10x^7$ (e) $\frac{6a}{b}$ (f) $4xy^3$
4 $n = 2$

25. Substitution

1 1 hour 36 min
2 (a) 46   (b) 9
3 (a) 33   (b) 37
4 (a) $18 - 4q$ (b) 40   (c) 45
5 (a) 27   (b) 28   (c) 8
6 2 cm

26. Formulae

1 6   **2** 602
3 (a) 6.7 m   (b) The gap between each rung.
4 96 kg   **5** 320 m
6 $3\sqrt{2}$ seconds or 4.24 (2 d.p.) seconds
7 By substitution of $f = 21$ and $g = -2$ it shows the left hand side equal to the right hand side. (By making a quadratic equation and solving, some students may realise that g could be equal to 4.5 also.)

27. Writing formulae

1 (a) $2M + 5N$ (b) £74.40
2 $M = N + 2$
3 (a) $V = 2qr^2$ (b) Surface area $= 2r(3q + 4r)$ or $6rq + 4r^2$
4 (a) $A + B + C = 20$
     (b) (i) $A = B + C$ (ii) $A = 2B + 2$

28. Expanding brackets

1 (a) $5x + 5$ (b) $6x - 3$ (c) $3 + x\sqrt{3}$
2 (a) $-2x - 6$ (b) $-8x - 4$ (c) $6x - 12$
3 (a) $x^2 + 3x$ (b) $4x^2 - 10x$ (c) $15xy + 10x^2$
4 (a) $6x + 12$ (b) $5x - 8$
5 $a = -3, b = 8$
6 $3A = 9x + 6$, $2B = 4x - 2$, $3A + 2B = 13x + 4$

29. Factorising

1 (a) $2(x + 4)$ (b) $12(x - 3)$ (c) $6(p - 5)$
2 (a) $x(x + 5)$ (b) $x(x - 14)$ (c) $a(a + 17)$
3 (a) $2x(x + 2)$ (b) $9r(3r - 1)$ (c) $8t(t - 8)$
4 (a) $4e(2e + 3)$ (b) $4x(3x - 4)$ (c) $7q(2q - 5)$
5 $L = 3x$, $W = x + 4$ (order not important)
6 If x is an odd number then $3x^2$ will also be odd as it does not have 2 as a factor. If y is odd also then for the same reason, $5xy$ will also be odd. When we add two odd numbers we always get an even result.

30. Linear equations 1

1 (a) $x = 6$ (b) $a = -5$ (c) $r = -20$
2 (a) $x = 14$ (b) $q = 5$
3 (a) $x = 7$ (b) $q = \frac{-16}{5}$ (c) $a = -1$
4 (a) $P = 8t + 10$ (b) 34 cm   (c) $t = 9$ cm

31. Linear equations 2

5 (a) $x = 3$ (b) $a = 8$ (c) $q = 3$
6 (a) $x = 3$ (b) $t = 2$ (c) $u = 1$
7 $n = 5$

32. Inequalities

1 (a) $x \geq 0$ (b) $x > 1$ (c) $-4 < x \leq 1$
2 (a) $x \leq 2$

     −5 −4 −3 −2 −1 0 1 2 3 4 5

   (b) $x > 2$

     −5 −4 −3 −2 −1 0 1 2 3 4 5

   (c) $-1 \leq x \leq 2$

     −5 −4 −3 −2 −1 0 1 2 3 4 5

3 (a) $-1, 0, 1, 2, 3$ (b) 13
     (c) $-3, -2, -1, 0, 1, 2$
4 (a) $35 \leq q < 45$ (b) $250 \leq r < 350$
     (c) $3.45 \leq u < 3.55$
5 ($t = $ time) $5h\,11.5m \leq t < 11h\,12.5m$

33. Solving inequalities
1. (a) $x \le 5$ (b) $x > 3$ (c) $x \le 0.5$
2. (a) $x \le 17$ (b) $x > 7$ (c) $x \le 4$
3. (a) $-5, -4, -3, -2, -1, 0, 1$ (b) $1, 2, 3$ (c) $1, 2$
4. $3, 4$

34. Sequences 1
1.

2. (a) 21, 26   (b) 14, 17   (c) 28, 36   (d) 64, 49
3. $-1, 3, 8, 14, 21, 29$
4. (a) $-1, -5, -9$ (b) Start at 15 and continue the sequence by subtracting 4 each time.
5. (a) 54   (b) 12, 18

35. Sequences 2
6. (a) $5n - 3$ (b) $4n - 1$ (c) $4n + 1$ (d) $3n + 8$
7. $t = 9n - 1$
8. (a) -9 and 39 (b) $12n - 21$
9. (a) $3n - 7$ (b) 38

36. Coordinates
1. (a) (i) $(2, 1)$ (ii) $(-1, 2)$

   (b) (i) (ii)

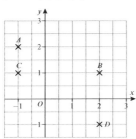

2. (a) $(9, 11)$ (b) $(5, 8)$
   (c) $(0, 10)$ (d) $(-3, -2)$
3. (a) $(0, 15)$ (b) $(10, 20)$

37. Gradients of lines
1. (a) 2   (b) $\frac{3}{4}$
2. (a) 4   (b) $\frac{1}{2}$
3. -1

38. Straight-line graphs 1
1. (a)
$y = 2x - 1$

x	-1	0	2
y	-3	-1	3

   (b)

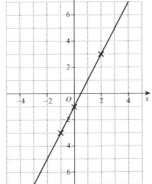

(c) No
(d) Yes, by substituting the x and y values in the coordinate into the equation.

2. $x - y = 4$

x	-12	0	12
y	-16	-4	8

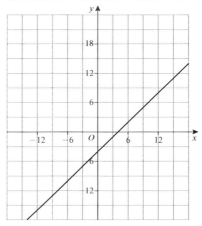

3. $y = x + 6$

39. Straight-line graphs 2
4. (a) $y = 2x - 1$ (b) $y = 5 - x$ or $x + y = 5$
5. $y = x + 2$
6. $y = -3x - 2$

40. Real-life graphs
1. (a) 48 km   (b) 38 miles
   (c) Bournemouth
2. (a) £120 000   (b) $36\,\text{m}^2$

41. Distance-time graphs
1. (a) 10:00   (b) 4 km
   (c) 15 minutes   (d) 8 km/h
2. (a)

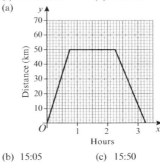

   (b) 15:05   (c) 15:50

42. Rates of change
1.

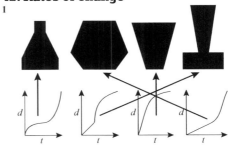

2. (a) 7000 litres   (b) 6
   (c) How many litres ($\times 1000$) per hour the pool is filling
3. (a) $2\,\text{m/s}^2$ (b) Constant velocity of 40 m/s
   (c) $-1\,\text{m/s}^2$

43. Expanding double brackets

1. (a) $x^2 + 3x + 2$ (b) $x^2 + 6x + 8$
   (c) $x^2 + 7x + 12$ (d) $x^2 + 2x - 3$
   (e) $x^2 - 5x + 6$
2. (a) $6x^2 + 7x + 2$ (b) $5x^2 + 7x - 6$
   (c) $9x^2 - 9x + 2$ (d) $x^2 + 10x + 25$
   (e) $4x^2 - 28x + 49$
3. When expanded it becomes $n^2 + 2n + 1 - n^2$. This simplifies to $2n + 1$. Any number described as $2n$ is even, as 2 is one of its factors. 1 is an odd number and any odd plus any even results in an odd number.
4. When expanded it becomes $a^2 + b^2 - a^2 + 2ab - b^2$. This simplifies to $2ab$. This implies that the result is an even number as 2 is a factor. Therefore the result can only be even.
5. It is true. See working based on the area of a triangle being $\frac{\text{base} \times \text{height}}{2}$:
   $$\text{Area} = \frac{(x + 2)(3x + 1)}{2}$$
   $$= \frac{3x^2 + x + 6x + 2}{2} = \frac{3x^2 + 7x + 2}{2}$$

44. Quadratic graphs

1. (a)

x	−3	−2	−1	0	1	2	3
y	8	3	0	−1	0	3	8

   (b)
   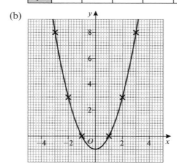
   (c) $(0, -1)$
   (d) Approximately 1.3

2. (a)

x	−3	−2	−1	0	1	2	3
y	12	5	0	−3	−4	−3	0

   (b)
   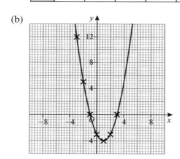
   (c) $(1, -4)$

45. Using quadratic graphs

1. (a)

x	−3	−2	−1	0	1	2	3
y	7	3	1	1	3	7	13

   (b)
   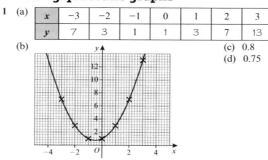
   (c) 0.8
   (d) 0.75

2. (a)

x	−4	−3	−2	−1	0	1	2
y	−4	1	4	5	4	1	−4

   (b)
   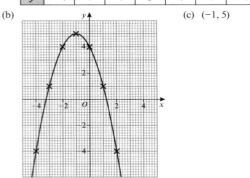
   (c) $(-1, 5)$

46. Factorising quadratics

1. (a) $4, -4$ (b) $\frac{1}{2}, -\frac{1}{2}$ (c) $6, -6$
2. (a) $x(x + 6)$ (b) $x(x + 12)$ (c) $x(x + 2)$
   (d) $x(x - 1)$ (e) $x(ax + b)$
3. (a) $(x + 1)(x + 3)$ (b) $(x - 3)(x + 4)$
   (c) $(x - 2)(x - 3)$
4. (a) $(x - 12)(x + 12)$ (b) $(x - 1)(x + 1)$
   (c) $2(x - 3)(x + 3)$

47. Quadratic equations

1. (a) $x = 0$ or $x = 3$ (b) $x = 0$ or $x = -3$
   (c) $x = 0$ or $x = 5$ (d) $x = 0$ or $x = -4$
2. (a) $x = 1$ or $x = -2$ (b) $x = 1$ or $x = 4$
   (c) $x = 3$ or $x = -7$ (d) $x = -3$ or $x = -6$
3. (a) $x = 4$ or $x = -4$ (b) $x = 10$ or $x = -10$
   (c) $x = 5$ or $x = -5$ (d) $x = -2$ or $x = 2$
4. $(x + 1)(x - 1) = 440$
   $x^2 - 1 = 440$
   $x^2 = 441$
   $\therefore x = 21$
   These workings show that the positive consecutive even numbers must be: $x + 1 = 22$ and $x - 1 = 20$

48. Cubic and reciprocal graphs

1. (a)

x	−2	−1	0	1	2
y	−1	3	1	−1	3

   (b)
   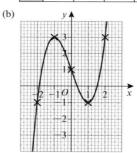
   (c) $(-1, 3)$ and $(1, -1)$
   (d) when $y = -3$ x is roughly -2.2
   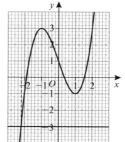

2. (i) e   (ii) a   (iii) c   (iv) d   (v) b

49. Simultaneous equations

1 (a) $x = 2, y = 1$ (b) $x = 5, y = -2$

2 when $y = -3$ x is roughly -2.2

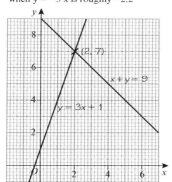

$x = 2, y = 7$

3

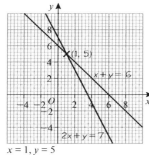

$x = 1, y = 5$

50. Rearranging formulae

1 (a) $a = \dfrac{c - 5}{3b}$

   (b) 3

2 $x = \dfrac{2y + 10}{3}$

3 $x = 4y - 7$

4 (a) $O = \dfrac{P - LW}{H}$

   (b) $O = 2$

51. Using algebra

1 (a) $12n - 15 = 345$

   (b) $n = 30$

2 $n = 70$

3 $y = 3x - 1$

4 $b = 4(3x + 2)$

52. Identities and proof

1 $(3n + 2)^2 + (3n - 2)^2$
$= 9n^2 + 6n + 6n + 4 + 9n^2 - 6n - 6n + 4$
$= 18n^2 + 8$
$= 2(9n^2 + 4)$

2 $n + (n + 1) = 2n + 1$
$2n$ is an even number as 2 is a factor. 1 is an odd number. Any even number plus any odd number results in an odd number.

3 $2(x + 1)(x + 2) = 2(x^2 + 3x + 2) = 2x^2 + 6x + 4$
If $2x^2 + 6x + 4 = Ax^2 + Bx + C$, then $A = 2, B = 6, C = 4$
Hence $A + B + C = 12$ no matter the value of x.

4 If $x = 20$, then $y = 2$. This implies, using the difference of 2 squares that $22 \times 18 = 20^2 - 2^2 = 396$

5 By factorisation we can see that both numbers have a factor 3:
$$\frac{(n + 2) + (n + 3) + (n + 4)}{n + (n + 1) + (n + 2)}$$
$$\frac{(n + 2) + (n + 3) + (n + 4)}{n + (n + 1) + (n + 2)} = \frac{3n + 9}{3n + 3} = \frac{3(n + 3)}{3(n + 1)}$$

53. Problem-solving practice 1

1 4 days

2 (a) $(6, 10)$ (b) $(-6, -6)$

3 (a) £755   (b) 2.5 hours

4 90–95 (inclusive)

5 (a) £10   (b) £8

54. Problem-solving practice 2

6 $n = 11$

7 $250\,187.5 = 250\,000$ (3 s.f.)

8 (a) $2n + 1$ (b) 35

9 $\dfrac{bh}{2} = \dfrac{(4x - 2)(3x + 8)}{2} = \dfrac{12x^2 + 26x - 16}{2}$
$= 6x^2 + 13x - 8$

10 (a) Perimeter = 32 cm
     (b) Area = 55 cm^2

RATIO & PROPORTION

55. Percentages

1 (a) 1.6   (b) 16.1

2 (a) 37.5%   (b) 85%

3 (a) £15.20   (b) £91.20

4 (a) 6.25%   (b) 37.5%
    (c) Mikita has a better estimate as Number and Algebra is $\frac{46}{80}$ of the questions which is 57.5%

56. Fractions, decimals and percentages

1 (a) $\frac{21}{50}$ (b) $\frac{14}{25}$ (c) $\frac{24}{25}$

2 (a) $0.6, \frac{2}{3}, 67\%$ (b) $38\%, \frac{29}{75}, 3.8$
    (c) $\frac{3}{20}, 0.151, \frac{4}{21}$

3 (a) 39   (b) 102

4 0.7 3 kg = 2.1 kg
0.6 4 kg = 2.4 kg
Nisha's birds ate the most.

57. Percentage change 1

1 (a) £82.80   (b) 44.8   (c) 29.16

2 (a) 25%   (b) 281.25%

3 (a) 46.4% (3 s.f.)   (b) 85.7% (3 s.f.)

4 (a) 95%
    (b) It is the same. Both salaries were scaled up by the same factor.

58. Percentage change 2

5 Ticket shop 2 is cheaper at a cost of £59.80 where Ticket shop 1 cost £61.71.

6 He is correct. A&P costs £140.80, Fixnfit costs £170.40.

7 Electrical City is the better deal at a cost of £174.24, Pasters is £177.60.

59. Ratio 1

1 (a) $1 : 5$ (b) $3 : 4$ (c) $12 : 25$

2 (a) $40 : 32$ (b) $15 : 27$ (c) $88 : 33$

3 (a) Tannu drove 95 miles, Millie drove 50 miles.
    (b) 22 miles

60. Ratio 2

4 (a) 1.4 kg   (b) 234 g

5 (a) $5 : 3$ (b) 312.5 kg

6 £8050   **7** 162   **8** 10

61. Metric units

1 (a) 3.5 cm   (b) 8.3 cm   (c) 420 mm
   (d) 2.3 kg   (e) 2 350 g   (f) 3.45 km

2 (a) 170 mm   (b) 18.5 cm   (c) 3.75 kg
   (d) 11 100 g   (e) 2300 ml   (f) 0.7551

3 6 glasses with 4 ml left over
4 44
5 2.4 g (2 s.f.)

62. Reverse percentages
1 £5400
2 £21 740
3 £235 000
4 Pete is wrong. A discount of 15% on £840 is £714 which is £126 less.
5 (a) Han   (b) £100

63. Growth and decay
1 £655.50
2 5 years
3 £12 550
4 (a) 1.5%   (b) £2703.58
5 No, it will have more than doubled, at the end of 5 years its worth will be £7464.96

64. Speed
1 30 m/s
2 66 km/h
3 432 km
4 8.5 (1 d.p.) mph over the limit
5 Bristol to Penzance at 44.4 mph

65. Density
1 $\frac{3}{1750\text{g/mm}^3}$
2 4.14 cm³ (2 d.p.)
3 Mass = 1775 g = 1.775 kg
4 54 g
5 (a) $r = 2\sqrt{5}$ cm $D = 2(2\sqrt{5}) = 4\sqrt{5}$ cm   (b) 7.75 g/cm³

66. Other compound measures
1 200 N/m²
2 21 420 N
3 5 minutes
4 $\frac{5}{3}$ N/cm²
5 (a) 0.75 litres (b) $\frac{3}{8}$ second

67. Proportion
1 £280
2 £8.25
3 21
4 25
5 6 days
6 £1500
7 The 900 g bag works out at 0.53…p/g. The 1.2 kg bag works out at 0.535 p/g.

68. Proportion and graphs
1 1260 N
2 20 m
3 $y = 1.17x$
4 (a) 38 N   (b) 10.4 cm
5 (a) $k = 300$ (b) 60 Pa

69. Problem-solving practice 1
1 French, mathematics, art
2 Close but actually 87.9% left for sale.
3 Sole2Sole is better at a total cost of £75.25 compared to Shooz cost of £78.
4 103 g
5 (a) 188.8 g   (b) 2 tin cubes and 1 brass cube

70. Problem-solving practice 2
6 24 300 litres
7 5-box works out at £1.18 exactly per box, therefore the best value.
8 21
9 (a) Hans has made £2775. 65
   (b) Nedra has £18 200.74
   (c) Yes, they have a total of £36 176.39

148

GEOMETRY & MEASURES

71. Symmetry
1 See diagram. O has an infinite number of lines of symmetry only 1 is shown.

2 No, it is untrue. A *regular* polygon with *n* sides has *n* lines of reflective symmetry.
3 (a) 1 line of symmetry

   (b) order 2 rotational symmetry

   (c) 2 lines of symmetry

   (d) order 3 rotational symmetry.

4 (a) Octagon shows possibilities

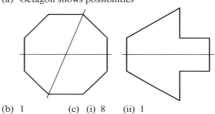

   (b) 1   (c) (i) 8   (ii) 1

72. Quadrilaterals
1

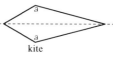

2 (a) One pair of parallel sides and one line of symmetry

isosceles trapezium

(b) Two pairs of parallel sides with unequal diagonals

(c) Only two equal obtuse angles

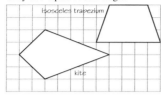

(d) A parallelogram which is not a square, a rectangle or a rhombus.

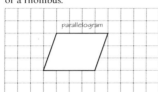

(e) Order 4 rotational symmetry

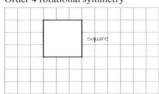

(f) Only two, equal acute angles

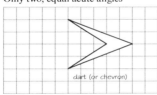

73. Angles 1
1 (a) obtuse   (b) acute
   (c) reflex   (d) right-angle
2 (a) acute (it is less than 90°)
   (b) obtuse (it is greater than 90°)
3 (a) 89°   (b) acute
4 (a) 30°   (b) 120°   (c) 150°

74. Angles 2
5 $a = 105°$, $b = 40°$, $x = 325°$
6 $a = 50°$, $b = 80°$, $x = 160°$
7 (a) 124°   (b) 124°   (c) 180°
   (d) No. $a = d$ still sum to 180°

75. Solving angle problems
1 (a)

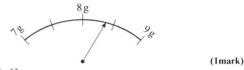

   (b) $d = a + b$ (c) $e = b + c$
2 $x = 11.9°$ (1 d.p.)
3 $x = 117°$

76. Angles in polygons
1 (a) 135°   (b) 45°
   (c) Interior angles of a triangle equal 180°.
     $6 \times 180° = 1080°$. Dividing this by the 8 equal angles at
     the vertices of the polygon gives 135°
2 (a) 24°   (b) 15
3 32.73°
4 108°

77. Time and timetables
1 (a) 3.45 pm   (b) 6.05 pm   (c) 10.09 am   (d) 9.59 am
2 (a) 19.15   (b) 22.47   (c) 04.03   (d) 01.42
3 34 h 40 m
4 10.09
5 (a) 1h 25 m   (b) 17.44   (c) 16.59

78. Reading scales
1 (a) 66   (b) 78   (c) 8.2   (d) 6.375
2 (a) 131   (b) 18.8

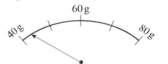

   (c) 3375   (d) 2175

3 8.5 g
4 (a) 8.3 g

(1mark)

   (b) 43 g

(1mark)

5 Llewellyn   (28.5 litres)

79. Perimeter and area
1 (a) 12 cm   (b) 5 cm²
2 (a) 11 cm²   (b) $\frac{11}{36}$
3 (a) 140 cm   (b) 80 cm
4 (a) 34 cm   (b) 42 cm²

80. Area formulae
1 (a) 28 cm²   (b) 32.5 cm²
   (c) 31.5 cm²
2 (a) 11.76 cm²   (b) 3 : 5
3 19 cm

81. Solving area problems
1 $x = 9$ cm
2 3.2 cm
3 16

82. 3D shapes
1 Pentagonal prism
2 (a) $n = 4$ cm   (b) $n = 6$ cm
3 336 cm²
4 (a) 268 cm²   (b) 14

83. Volume of cuboids
1 (a) 308 cm³   (b) 360 cm³   **(c)** 660 cm³
2 152.72 cm (2 d.p.)
3 80 cm

84. Prisms
1 (a) 360 cm³   (b) 440 cm³   (c) 660 cm³
2 (a) 64 cm   (b) 180 cm²   (c) 144 cm³

85. Units of area and volume

1 (a) $400\,\text{mm}^2$ (b) $3\,000\,000\,\text{m}^2$ (c) $8\,\text{cm}^2$
2 (a) $4\,000\,000\,\text{cm}^3$ (b) $30\,000\,\text{mm}^3$ (c) $0.25\,\text{cm}^3$
3 (a) 2.7 litres   (b) 9000 litres   (c) 17 000 litres
4 (a) 0.96 litres   (b) 270 000 litres

86. Translations

1 (a) $\binom{4}{5}$ (b) $\binom{7}{-3}$ (c) $\binom{-1}{3}$ (d) $\binom{-4}{0}$

2 (a)

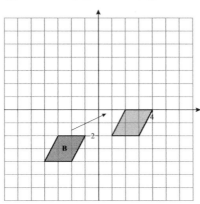

(b)

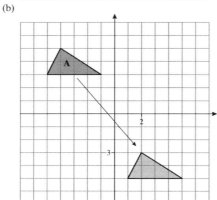

3 (a) Translation through vector $\binom{-1}{6}$ (b) $\binom{-5}{-4}$

87. Reflections

1 (a)

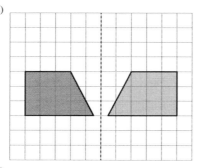

(b)

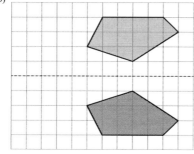

(c)

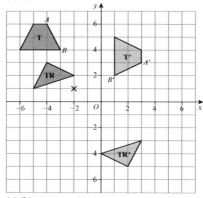

2

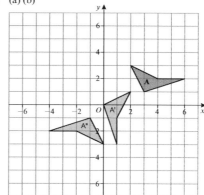

3 (a) Translation in $\binom{7}{-7}$ (b) Translation in $\binom{-5}{2}$

88. Rotations

1 (a) (b)

2 (a) (b)

3 (a) Rotation 90° clockwise about (1, 2)
   (b) Rotation 180° about (2, 1)

89. Enlargements

1 3
2 (a) $\frac{2}{3}$ (b) 13 cm
3 Scale factor enlargement 3, centre of enlargement (−5, 5)

90. Pythagoras' theorem

1 (a) $h = 8.3$ cm (1 d.p.) (b) 4.8 cm (1 d.p.)
   (c) 10.2 cm
2 17.3 cm (1 d.p.)
3 119.8 km (1 d.p.)
4 No it does not as the bottom litter bin is 35.77 m away from its nearest one.

91. Line segments

1 9.4 (1 d.p.)
2 13.3 (1 d.p.)
3 5.4 (1 d.p.)
4 (a) 49 km (b) 28 km (c) 56.4 km (1 d.p.)

92. Trigonometry 1

1 (a) 12.4 cm (1 d.p.) (b) $a = 8.2$ cm (1 d.p.)
2 4.7 m (1 d.p.)
3 1.77 m (2 d.p.)
4 37.23 m (2 d.p.)

93. Trigonometry 2

5 Both are right-angled triangles with the same angle θ therefore the remaining angle must be the same on both triangles too and hence the sides will be in the same ratio.
6 (a) 48.2° (b) 51.0°
7 (a) θ = 27.6° (1 d.p.)
   (b) (i) 6.72 cm (2 d.p.)
       (ii) evidence of $q^2 + (12.8511...)^2 = 14.5^2$
8 10.2°

94. Solving trigonometry problems

1 (a)
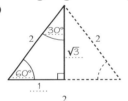

b
	0°	30°	45°	60°	90°
sin	O	$\frac{1}{2}$	$\frac{\sqrt{2}}{2}$	$\frac{\sqrt{3}}{3}$	1
cos	1	$\frac{\sqrt{3}}{2}$	$\frac{\sqrt{2}}{2}$	$\frac{1}{2}$	O
tan	O	$\frac{\sqrt{3}}{3}$	1	$\sqrt{3}$	Not defined

2 $y = \frac{\sqrt{5}}{2}$ cm
3 (a) 57.7° (b) 49.1°
4 (a) 10.6 cm (3 s.f.) (b) 18.4 cm

95. Measuring and drawing angles

1 (a) 64° (b) 65°
   (c) (i) 295° or (ii) 245°
2 (a)

   (b)
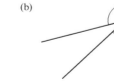
3 (a) 135° (b) 324°
4 Check both angles given then 3 sides.

96. Measuring lines

1 All sides must be measured by marker.
2 Marker to check each length drawn.

3 A ——————————————|——————— B
4 23 m
5 $1\frac{2}{3}$ m

97. Plans and elevations

1 (a) cylinder (b) square-based pyramid
   (c) triangular prism
2 (a)
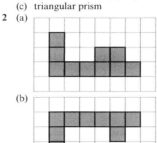
   (b)
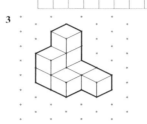
3

4 Side
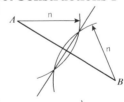

98. Scale drawings and maps

1 (a) 3 m (b) 5.4 km (c) 1.788 km
2 (a) 16.2 km (b) 9.7 km (c) 40 km
3 Model will be 3.9 m long and therefore would fit inside a 4 m room.
4 0.1 m/s

99. Constructions 1

1

2
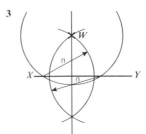
3

100. Constructions 2

4 Check all sides are 5 cm. Angles not important.

5

6

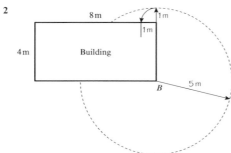

7

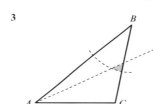

101. Loci

1

2

3

4

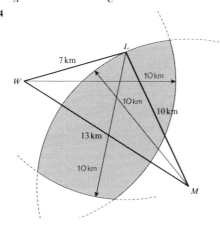

102. Bearings

1 (a) (i) 067°   (ii) 293°   (b)   (i) 108.5°   (ii) 288.5°
2 (a) Measure 36°.   (b)   Measure reflex angle 185°.
3 (a) 078°

(b)

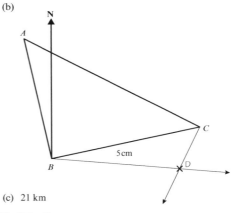

(c) 21 km

103. Circles

1

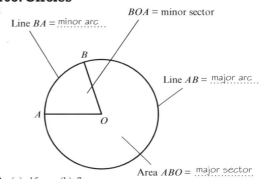

Line *BA* = minor arc   *BOA* = minor sector

Line *AB* = major arc

Area *ABO* = major sector

2 (a) 15π (b) 7π
3 (a) $r = \frac{27}{2\pi}$ cm

 (b) $r = 45\pi$ m
4 (a) $28 + 7\pi$ (b) $7.52 + 3.6\pi$
5 70.7

104. Area of a circle

1 (a) 81π mm²   (b) 81π m²
2 (a) 100π cm²   (b) 32π m²   (c) 768π mm²
3 (a) 1180 mm²   (b) 1.72 cm²

105. Sectors of circles

1 (a) $\frac{22}{5}\pi$ cm²   (b) $\frac{45}{4}\pi$ cm²
2 (a) $(4\pi + 12)$ cm

 (b) $\left(\frac{25}{2}\pi + 15\right)$ mm
3 (a) 9.01 mm²   (b) $\frac{2645\pi}{8}$ cm²

106. Cylinders

1 (a) 283 mm²   (b) 81.7 cm²   (c) 127 cm²
2 (a) 72π cm²   (b) 57π cm²   (c) 430π cm²
3 (a) 0.764 cm   (b) Cylinder

107. Volumes of 3D shapes

1 (a) $\frac{250\pi}{3}$ cm³   (b) $\frac{575\pi}{3}$ cm³   (c) 300π cm³
2 (a) $\frac{2048\pi}{3}$ cm³   (b) 360π cm³   (c) 441π cm³
3 40 cm

108. Surface area

1 $n = 4$ cm
2 $n = 10$ cm
3 $\frac{147\pi}{2}$ cm²
4 $h = 9.5$ cm
5 (a) 758.45 cm² (b) 27.20 m²
6 (a) $3\pi r^2$ (b) 1425.87 cm²

109. Similarity and congruence

1 (a) A and D   (b) B and C

2 (a)

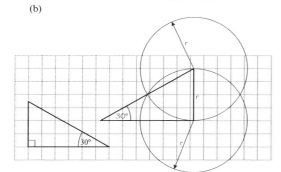

(b)

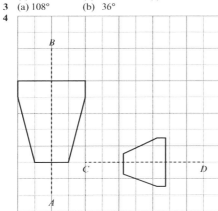

3 If both $a = 10$ mm and $x = 10$ mm then all three rectangles are congruent. The values are not given so this cannot be said. However, all three rectangles are similar.

110. Similar shapes

1 (a) 60°
  (b) $p = \dfrac{20\sqrt{3}}{3}$
  (c) $q = p = \dfrac{20\sqrt{3}}{3}$
  (d) $110\sqrt{2}$ cm

2 (a) 27°   (b) 45°   (c) 6.27 cm

3 (a) 108°   (b) 36°

4

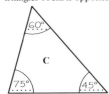

111. Congruent triangles

1 Missing angle on left is 52°. On right $52° + 16\theta = 180°$ therefore $\theta = 8°$ (giving unknown angles 72° and 56°). On both triangles 10 cm is opposite 56° hence triangles are congruent.

2

3 Yes, it is possible. As there is an unknown angle at a point made by two unknown sides, the unknowns on one triangle can be different to those on the other.

112. Vectors

1

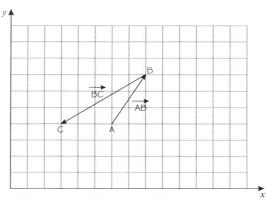

2 (a) $\begin{pmatrix} 8 \\ 2 \end{pmatrix}$ (b) $\begin{pmatrix} 2 \\ -4 \end{pmatrix}$ (c) $\begin{pmatrix} 5 \\ 4 \end{pmatrix}$
  (d) $\begin{pmatrix} 5 \\ -6 \end{pmatrix}$ (e) $\begin{pmatrix} 10 \\ -2 \end{pmatrix}$ (f) $\begin{pmatrix} -3 \\ -8 \end{pmatrix}$

3 (a) $\mathbf{q} + \mathbf{r}$ (b) $-\mathbf{q} - \mathbf{r}$ (c) $\mathbf{q} - \mathbf{r}$ (d) $\mathbf{r} - \mathbf{q}$

4 (a) $3\mathbf{a}$ (b) $-3\mathbf{a}$

113. Problem-solving practice 1

1 (a) $x = 60°$ (b) $y = 110°$ (c) $z = 20°$

2 Volume of cylinder A = 216π, volume of cylinder B = 108π hence A = 2B

3 Reflection in $y = 0$

114. Problem-solving practice 2

4 130°

5 73.1 (3 s.f.) km

6 2.84 cm (3 s.f.)

PROBABILITY & STATISTICS

115. Two-way tables

1

	Berlin	Bielefeld	Bremen	Total
Football	8	11	12	31
Tennis	22	3	9	34
Total	30	14	21	65

2

	Printing	Painting	Pottery	Total
Girls	20	17	7	44
Boys	22	16	2	40
Total	42	33	9	84

3 (a)

	Europe	UK	Neither	Total
Under 30	29	5	3	37
30–55	31	14	19	64
Over 55	28	0	21	49
Total	88	19	43	150

(b) 59

(c) The survey only specified holidays in Europe or UK. America, Asia and Africa for example could be in the neither column.

116. Pictograms

1 (a) True because the difference is equal to one house on the pictogram which is half shown in May.
  (b) 36 houses (c) 150 houses

2 (a) 40 visitors = 1 car   (b) 220
  (c) Saturday = 2.5 cars, Sunday = 1.5 cars

117. Bar charts

1 (a)

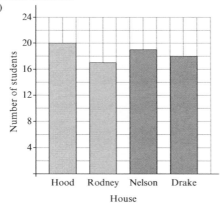

(b) False: The bars are not equal but they are close in value to each other and random sampling does not guarantee a theoretically expected result.

2 (a) (i) (ii)

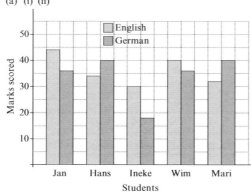

(b) Hans and Mari (c) Ineke

118. Pie charts

1

 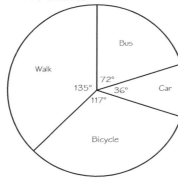

2

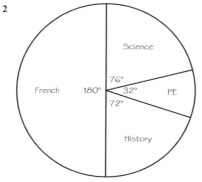

3

County	Frequency
Lancashire	180
Sussex	328
Somerset	92
Surrey	120

119. Scatter graphs

1 (a) Positive

(b)

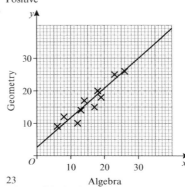

(c) 23

(d) It is possible, but it would be an anomaly as the correlation is so strong.

2 (a) No correlation

(b) It is not possible as there is no correlation between distance run and money raised.

(c) His reasoning is spurious as this implies that there is correlation in the data collected.

120. Averages and ranges

1 (a) 9 (b) 9 (c) 9 (d) 11

2 (a) £35 400 (b) £32 800

3 8, 8, 9, 13, 16

121. Averages from tables 1

1 (a)

C	f	Cf
0	4	$0 \times 4 = 0$
1	7	7
2	10	20
3	6	18
4	5	20
$\Sigma(f)$	32	65 $\Sigma(Cf)$

(b) 2 countries

(c) 2 countries

(d) 2.03 (3 s.f.) countries

(e) 4

2 (a)

V	f	Vf
0	8	0
1	5	5
2	6	12
3	4	12
4	3	12
5	1	5
$\Sigma(f)$	27	46 $\Sigma(Vf)$

(b) He has taken the largest frequency. He should have claimed 0 was the mode as this was seen 8 times.

(c) Mode = 0, Median = 2, Mean = 1.7 (2 s.f.)

(d) She is correct as the median position would be at position 14.5 which is still 2.

122. Averages from tables 2

3 (a)

Number of hours (*h*)	Frequency (*f*)	midpoint (*x*)	*fx*
$0 \leqslant h < 2$	1	1	$1 \times 1 = 1$
$2 \leqslant h < 4$	5	3	15
$4 \leqslant h < 6$	6	5	30
$6 \leqslant h < 8$	3	7	21
$\Sigma(f)$	15		67 $\Sigma(fx)$

(b) $4 \leqslant h < 6$ (c) $4 \leqslant h < 6$ (d) 4.5h (e) 8h

4 (a)

Money (€)	*f*	mid	f(mid)
$0 \leqslant € < 10$	4	5	20
$10 \leqslant € < 20$	6	15	90
$20 \leqslant € < 30$	8	25	200
$30 \leqslant € < 40$	3	35	105
$\sum(f)$	21	$\sum f(mid)$	415

(b) €19.76

(c) This will take the estimated mean to €20. The median before Méganne added her data was to be found in the class interval $20 \leq € < 30$. After she added her data the median was to be found in the same class interval. As we do not know the individual data entered it is not possible to know the exact value of the median in either. However, the median will be greater after she added her data.

123. Line graphs

1 (a)

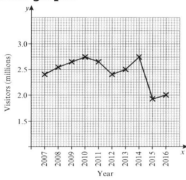

(b) Gentle upward trend at the start but followed by a steeper downward trend with a slight recovery at the end.

2 (a) It is a 31 day month. June has only 30 days.

(b) The median is 2 (the 16th entry).

3 (a) 5

(b) The mean shoe size would be 2.41 (3 s.f.). No such shoe size exists.

124. Stem-and-leaf diagrams

1 (a)

```
5 | 0
4 | 1 2 2 4 9
3 | 0 1 2 7 8 8 9
2 | 5 7 8 9
1 | 1 9 9
0 |
```

Key: 5 | 0 = 50

(b) 42, 38 and 19 are all modal scores

(c) 33.55

(d) 39

2 (a)

```
12 | 4
11 | 5
10 | 3 5
 9 | 1 2 5 7 7
 8 | 6 7
 7 | 1 2
 6 | 1 8
 5 | 3
 4 | 1 9
 3 | 4
 2 | 5
```

Key: 2 | 5 = 25

(b) 97 (c) 87 (d) 99

3 (a) The number 1 appears 3 times in one leaf which is more than any other number does.

(b) £51 (c) £44.50 (d) $x = 3$

125. Sampling

1 (a) £45 (b) £40

(c) He surveyed his friends, which is not a good sample from the class so the estimations may not be very close to reality.

(d) Choose a sample of classmates by giving them all a number and then put numbers in a hat. The same could be done by generating random numbers on a calculator to match classmates' numbers.

2 (a) 15s (b) 50cm

(c) Change measurements to mm.

(d) The length of lower legs is not infinite and there is a limit to the minimum time to complete it.

126. Stratified sampling

1 (a) 85 (b) 149

2 (a) 320 female workers

(b) 120 male office workers

3 5

127. Comparing data

1 Mean score in German is greater than Italian with a shorter range too.

2 Although the mean is 1 point better in Statistics than Algebra, the range is half that in Algebra. Algebra should be seen as the subject that is doing better.

3 (a) The 10.5th position shows that the median number of pupils absent each week was 1. On average over time there was only one pupil absent each day over the 20 day period, this is possible because of the high number of days recorded when no pupil was absent.

(b) 4 absent pupils

(c) Mrs Wright's class has a better record as both median and range are lower than Mr Patel's class.

4 The mean scores (to 2 s.f.) are the same with 66 for both schools. However the range of scores is much lower for BH at 29 compared to that of 43 for MA and BH has a higher median than MA. The students at BH should be seen as performing better in English.

128. Probability 1

1 (a) Tuesday will follow Monday

(b) A random person has a birthday in November

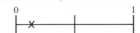

(c) A dice roll scores lower than 3

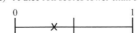

(d) Two fair coins both land on heads

Probability of heads = $\frac{1}{2}$

$\frac{1}{2} \times \frac{1}{2} = \frac{1}{4}$

2 (a) $\frac{1}{3}$ (b) $\frac{1}{6}$ (c) $\frac{1}{3}$ (d) 0
3 (a) even chance   (b) certain   (c) impossible

129. Probability 2
4 (a) 0.68   (b) 0.32
5

Number	0	1	2	3	4	5	6
Probability	0.25	0.25	0.15	0.1	0.1	0.1	0.05

6 (a) 0.48   (b) 0.22   (c) 0.125   (d) 0.3

130. Relative frequency
1 (a) 0.063   (b) 0.73
2 (a) 0.35   (b) 0.35   (c) 0.05
3 (a) 0.8
   (b) It is an estimate based on previous results, it does not take into account improvements over the period or any drop in form.
   (c) 7

131. Frequency and outcomes
1 $\frac{1}{6}$
2 (a)

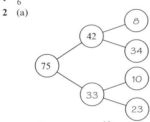

   (b) $\frac{19}{25}$ (c) $\frac{10}{33}$
3 (a) $\frac{7}{10}$ (b) 6

132. Venn diagrams
1 (a) Guests who chose bacon **and** egg
   (b) 14   (c) 25   (d) 3
2 (a)

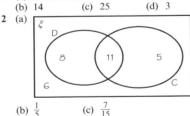

   (b) $\frac{1}{5}$ (c) $\frac{7}{15}$

133. Independent events
1 (a) $\frac{1}{9}$ (b) $\frac{1}{4}$ (c) $\frac{1}{3}$
2 (a)

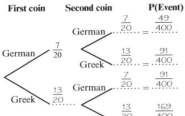

   (b) $\frac{351}{400}$

3 (a)

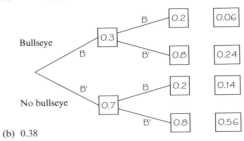

   (b) 0.38

134. Problem-solving practice 1
1 (a)

	Chinese	Indian	Italian	Total
Male	7	12	14	33
Female	9	10	8	27
Total	16	22	22	60

   (b) $\frac{6}{11}$
2 (a)

	Badminton	Pottery	Spanish	Guitar	Chess
Probability	0.3	0.1	0.15	0.4	0.05

   (b) 0.9   (c) 26

3 Woodwork recorded a median estimation of 6.1 cm and mathematics scored 6.0. Both classes showed a majority cluster around their respective medians but the range for woodwork was 24 and that of mathematics, 28. The woodwork class have slightly better estimation skills.

135. Problem-solving practice 2
4

Response	Frequency	Angle
Happy	6	24°
Unhappy	15	60°
Undecided	69	276°
Total	90	

5 (a)

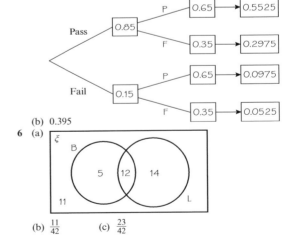

   (b) 0.395
6 (a)

   (b) $\frac{11}{42}$ (c) $\frac{23}{42}$

PRACTICE PAPER

Practice paper 1
1 (a) Baku at 28 m   (b) 855 m
   (c) 4 m   (d) 3427 m

2 (a) November sales 6893
October profit 2413
November costs 4318
December sales 7276
Total costs 17522
Total sales 28 605 Total profit 10 981
   (b) £7276 (c) £4100 (d) £652
   (e) £10 981 (f) £2745.25

3 (a) (i) It represents $\frac{5}{12}$ which is the highest single probability.
   (ii) 0.25
   (b) $\frac{25}{144}$

4 (a) 30 km
   (b) Tristan's journey is 1 km shorter.
   (c) No. In the first 3 triangles the ratio $H : V = 4 : 3$ and the new triangle has a ratio $H : V = 5 : 4$

5

	Supersave	Quickshop	Ecoworld	Other
Female	15	17	13	9
Male	17	18	12	7

6 (a) $2x$ (b) $12a$
   (c) $5x + 7y$ (d) $6x^2 - 4x - 2$

7 (a) $30^2 - 3^2 = 891$
   (b) $10^2 - 0.5^2 = 99\frac{3}{4}$

8 $\frac{1}{3}$ $\frac{3}{8}$ $\frac{2}{5}$ $\frac{3}{7}$ $\frac{4}{9}$

9 (a) isosceles (b) $r = 80°$ (alternate angles), $p = 80°$ (alternate angles), $q = 20°$

10 1 : 200 000

11 (a) $4(3x - 2y + 4)$
   (b) $5x(x^2 - 5x + 3)$

12 (a)

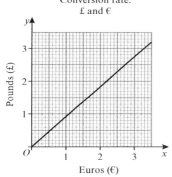

Conversion rate: £ and €

   (b) £25.50 (c) €600

13 194

14 (a) £6069 (b) 72%

15 (a)

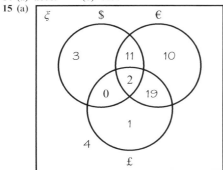

   (b) $\frac{14}{25}$

16 139 cm²

17 $21\frac{3}{7}$ km

18 (a) 18
   (b) The nth term of the sequence is defined as $t = 4n - 10$. If $t = 260$ then $260 = 4n - 10$ which implies $4n = 270$. If we divide both sides by 4, $n = 67.5$, which is not a whole number.

19 (a) 34 km (b) 5.23×10^5 mm

20 94.5°

21 (a)

$$2x - y = 8$$

x	0	2	4
y	-8	-4	0

$$x + 2y = -1$$

x	-3	1	3
y	1	-1	-2

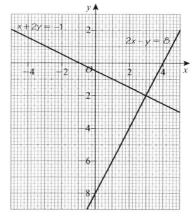

   (b) $x = 3,\ y = -2$

22

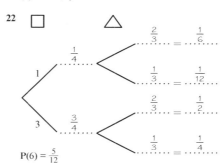

$P(6) = \frac{5}{12}$

Published by Pearson Education Limited, 80 Strand, London, WC2R 0RL.

www.pearsonschoolsandfecolleges.co.uk

Copies of official specifications for all Pearson qualifications may be found on the website: qualifications.pearson.com

Text and illustrations © Pearson Education Ltd 2018
Typeset and illustrated by York Publishing Solutions Pvt Ltd, India.
Commissioning, editorial and project management services by Haremi Ltd.
Cover illustration by Miriam Sturdee

The right of Russell Timmins to be identified as author of this work has been asserted by him in accordance with the Copyright, Designs and Patents Act 1988.

First published 2018

21 20 19 18
10 9 8 7 6 5 4 3 2 1

British Library Cataloguing in Publication Data
A catalogue record for this book is available from the British Library

ISBN 978 1 292 21371 2

Printed in Italy by L.E.G.O.

Notes from the publisher
1. While the publishers have made every attempt to ensure that advice on the qualification and its assessment is accurate, the official specification and associated assessment guidance materials are the only authoritative source of information and should always be referred to for definitive guidance.
Pearson examiners have not contributed to any sections in this resource relevant to examination papers for which they have responsibility.
2. Pearson has robust editorial processes, including answer and fact checks, to ensure the accuracy of the content in this publication, and every effort is made to ensure this publication is free of errors. We are, however, only human, and occasionally errors do occur. Pearson is not liable for any misunderstandings that arise as a result of errors in this publication, but it is our priority to ensure that the content is accurate. If you spot an error, please do contact us at resourcescorrections@pearson.com so we can make sure it is corrected.

Printed in Great Britain
by Amazon